Analog Devices and Circuits 1

Analog Devices and Circuits 1

Analog Devices

Christian Gontrand

WILEY

First published 2023 in Great Britain and the United States by ISTE Ltd and John Wiley & Sons, Inc.

ISTE Ltd
27-37 St George's Road
London SW19 4EU
UK

www.iste.co.uk

John Wiley & Sons, Inc.
111 River Street
Hoboken, NJ 07030
USA

www.wiley.com

Any opinions, findings, and conclusions or recommendations expressed in this material are those of the author(s), contributor(s) or editor(s) and do not necessarily reflect the views of ISTE Group.

Library of Congress Control Number: 2023942763

British Library Cataloguing-in-Publication Data
A CIP record for this book is available from the British Library
ISBN 978-1-78630-899-3

Contents

Preface

At the end of the Second World War, a new technological trend was born: integrated electronics. The latter relied on the enormous rise of integrable electronic devices; this field has invaded our societies.

In fact, electronics dates back to the beginning of the 20th century. Lee de Forest invented the triode in 1907. However, the first reliable tubes did not appear until the middle of the last century.

The frequency of these tubes increased in the late 1950s.

Solid-state physics, apart from galenic detectors and dry rectifiers, began at the beginning of the 1930s, with regard to the theoretical aspects (see Bloch (1929) and Wilson (1931)). At the end of December 1947, John Bardeen, Walter Houser Brattain (founders) and William Bradford Shockley (theorist, somewhat "founded" elsewhere) presented the first bipolar transistor to their colleague at (Ma) Bell Telephone in 1947. Over the past 20 years, huge technological advances have improved performance in terms of frequency, power and heat, with cost prices falling concomitantly. Then, links between research, development and productivity became very strong, with physicists and technologists tending to be confused for the other, both being material specialists, and the device, the circuit and the system forming the same ad hoc body. In the 1970–1980 decade, another field – numerical modeling and simulations – appeared; at the beginning of this decade, numerical mathematicians were curious guests at "brainstorming" meetings. At the end of this decade, they became full-fledged collaborators. In this case, with modeling being reliable and simulations robust, industrialists quickly realized their advantages, particularly in terms of cost, avoiding, among other things, the need to build demonstrators, to "miss" some technological step, leading to a whole series of wafers becoming "rubbish".

Moreover, on the pedagogical side, it contributes a lot to the understanding of physical and electrical phenomena, making it possible to carry out "numerical experiments", avoiding "breakage" in real devices and circuits. A new field has emerged in recent years: artificial intelligence, making it possible to make discoveries, often in heuristic form, using a large amount of data. Currently, we can talk about vertical artificial intelligence (AI), improving, refining, some diagnosis. It is surely "horizontal" AI that will revolutionize our lives. Robots or software will surely succeed in linking phenomena that seem, a priori, disjointed, perhaps succeeding in emulating human reasoning, or even the famous intuition that arises in the brains of the greatest scientists. We may then be inclined to accept that the theory of Charles Darwin (and Alfred Russel Wallace) continues to evolve, especially in our field of interest. But that is another story…

This book consists of two volumes: one deals with analog devices and the second deals with associated analog circuits. It is the substrate of two master courses, for example, DEA – Advanced Studies Degree (current Master 2) – taught under three establishments: Ecole Centrale, Institut National des Sciences Appliquées (National Institute of Applied Sciences), Chemistry, Physics, and Electronics (Chemistry and Digital Sciences), of the University of Lyon. Former students, having since set up shop in the industry, asked to transform related handouts into a book, in particular because the current training perhaps favors the qualitative over the quantitative too much. Note, of course, the existence of sums, most of them by yankees. Therefore, the aim here is not to carry out an exhaustive processing, but rather to work with key points, highlighting the complexity of the field, by focusing on certain major points, with a necessity of compromise which makes the field of microelectronics all the richer and more exciting. A lecturer post was then created at the end of the 1980s, because of the creation of the Centre Interuniversitaire de MIcroélectronique de la Région LYonnaise (CIMIRLY).

I would like to thank some of my colleagues, most often former doctoral students, master's students, or fifth year students of engineering; the list is not exhaustive: Saïda Latreche, Maya Lakhdara, Samir Labiod, Bruno Villard, Iulian Gradinariu, Fengyuan Sun, Anne Gérodolle, Serge Martin, Daniel Mathiot , Alain Chantre, Pascal Chevalier, Bruno Villard, Daniel Barbier, Drissi Fayçal, Filali Omar, Marc Buffat, Francis Calmon, Jacques Verdier, Pierre-Jean Viverge, Mohamed Abouelatta, Yue Ma, Christian Andrei, Olivier Valorge, Florent Miller, Rabah Dahmani, Rachid Benslimane, Alain Poncet, Michel Le Helley, Jean-Pierre Chante, Geoffroy Klisnick, Jean-Claude Vaissière, Daniel Gasquet and Jean-Pierre Nougier.

Acknowledgments

The author would like to thank the Université de Lyon, Institut National des Sciences Appliquées, Union for the Mediterranean (UfM), and Euromed University of Fes (UEMF) for supporting the study.

August 2023

Introduction

I.1. Synoptic history of microelectronics

I.1.1. *Electricity: Ampere, Coulomb, Faraday, Gauss, Henry, Kirchkoff, Maxwell and Ohm*

– 1826: Ohm's law (G.S. Ohm);

– 1837: S. Morse (New York) → Telegraph: binary signals: dot-dash:

  - W. Thomson and C. Wheastone;

– 1865: J.C. Maxwell → Electromagnetism:

  - H. Hertz → Production of electromagnetic waves in the laboratory;

– 1876: A.G. Bell → Telephone;

– 1877: T. Edison → Phonograph (disc: First ROM);

– 1996: G. Marconi → Wireless phone: radio waves (~ km).

I.1.2. *Vaccum tube*

– 1895: H.A. Lorentz → Electron (< Greek: amber); discrete charges;

– 1897: J.J. Thomson → Experiment → Existence of electrons:

  - K. Braun: cathode ray tube; first electron tube;

– 1904: A. Fleming → Invention of the diode (tube) → detector;

For a color version of all figures in this chapter, see http://www.iste.co.uk/gontrand/analog. zip.

– 1905: A. Einstein, H.A. Lorentz, H. Poincaré → Special relativity: intrinsic to electromagnetism;

– 1906: G.W. Pickard → Silicon crystal (Si) detector with whiskers:

- poor reliability because of spikes;

– 1906: L. de Forest → Audion triode (diode + gate: ancestor of the transistor): first controlled source.

Initial applications

– 1884: A.I.E.E: American Institute of Electrical Engineers;

– 1906: A.I.E.E + I.R.E → I.E.E.E: Institute of Electrical and Electronics Engineers;

– 1911: the triode is reliable (cathode covered with an oxide layer + very high vacuum):

- telephony and radiocommunications;

– 1917: creation of the Institute of Radio Engineers (I.R.E.).

Diodes and triodes

– 1912: E.H. Armstrong: feedback amplifier, cascading:

- L. de Forest: oscillator;

– 1917: E.H. Armstrong: heterodyne (see frequency translation);

– 1918: W. Eccles – F.W. Jordan: multivibrators:

- positive reaction, cascade amp + heterodyne → Detection of "weak signals";

– 1928: J.E. Lilienfeld: patents on the bases of field-effect transistors;

– 1930: E.H. Armstrong: frequency modulation (FM), before amplitude modulation (AM);

– 1930: B&W TV (black and white);

– 1942: Radar (RAdio Detection and Ranging):

- microwaves: Klystron, magnetron, etc.

I.1.3. *Computers (transistors – trans-resistors - integrated circuits – IC)*

– 1633: W. Schickard: MECHANICAL COMPUTER (wheeled, with different number of spokes);

– 1643: B. Pascal (see Pascaline);

– 1687: G.W. Leibnitz;

– 1883: C. Babbage: "The Analytical Machine":

  - perforated cards (see 1853; Jacquard): recorded programs;

– 1930: Mark I: H. Aiken (Harvard: 1930): automatic computer, with programmed sequences: ~17 m*3;

– 1936: A. Turing: general principles of automatic state machines;

– 1945: IBM: Industrial Business Machines; the 603: commercialized (701 in 1952, 704 in 1954 (144 kb memory));

– 1945: J. Von Neumann: theory on the architecture of automatic computers;

– 1946: J.P. Eckert, J. Mauchly (Pennsylvania):

  - E.N.I.A.C (Electronical Numerical Integration and Computer),

  - army $\rightarrow$ ballistic; J. Von Neumann $\rightarrow$ binary,

  - 40 × 2,300 tubes (10 m × 13 m space);

– 1947: IBM 604: 4,000 units in 12 years;

– 1948: the beginning of the computer industry; advent of the transistor (trans(RES)istor);

– 1951: UNIVAC 1: the first commercial computer;

– 1954: IBM 650: first-generation digital computers:

  - V. Bush (Massachusetts Institute of Technology); differential Analyzer (first electromechanical analog computer),

  - operational amplifier: analog electronics;

– 1955: first computer network: SABRE (created for American Airlines):

  - W. Shockley left Ma Bell to found his own company in Palo Alto, California, the first in what would become Silicon Valley;

– 1956: S. Cray: founded Control Data Corporations and later Cray Computers:

  - semiconductor computers;

– 1957: J. Backus: the first "superior" programming language: FORTRAN (Formula Translation);

– 1959: IBM 5090/7094: second generation:

  - PDP 1: the first interactive computer of the Digital Equipment Corporation,

- PDP8 (1965): first minicomputer in industry;

– 1964 IBM 360; with hybrid integrated circuits (ICs) with discrete transistors on a (bulk) substrate:

- Burroughs Control DATA, UNIVAC;

– 1970: IBM 370: third generation;

– 1980: fourth generation (CI very large scale integration (VLSI)):

- several tens of millions of operations/seconds,

- new architectures (vector, pipeline);

– 2010: Teles Hexaflop Computers.

I.1.4. *Analysis and theory*

It includes circuit analysis and synthesis techniques.

– BELL & MIT;

– H. Bode, H. Nyquist: feedback amplifiers:

- C.E. Shannon, V.A. Kotelnikov, A. Spataru: information theory (data transmission),

- for example, MIC: Reeves pulse code modulation;

– C.E. Shannon (1937) Boolean algebra → Analysis and design of switching circuits;

– A. Turing: universal computer concept;

– M.V. Wilkes: microprogramming;

– J.R. Raggazini, L.A. Zadeh: sampled data systems → Digital command control.

I.1.5. *Transistor*

– 1930–1945: study of the electromagnetic properties of semiconductors and metals:

- Block, Davidov, Lark, Horovitz, Mott, Schottcky, Slater, Summerfield, Vanvleck, Wigner, Wilson, Van der Ziel, Van Vliet;

– End of 1947 (J. Bardeen, W. Brattain, W. Shockley): the bipolar transistor and BELL telephone;

– 1950: Team (Bell labs): AT&T Research Branch:

- drawing (Czocralski method) ultrapure single crystals of germanium (Ge);

– 1951: commercial production of transistors:

- ATT provides patent licenses for transistor manufacturing → RCA Raytheon, General Electric, Westinghouse, Western Electric (ATT's manufacturing arm);

– 1954: Texas Instrument (<Teal): production of silicon transistors: Si;

– 1956: (J. Bardeen, W. Brattain, W. Schockley): Nobel Prize;

– 1975: ESAKI: band gap heterojunction engineering.

I.1.6. *Integrated circuits*

– 1958/59: J. Kilby, Texas Instruments; IRE "Semiconductor Circuit" Congress (multivibrator oscillators, Si or Ge capacitors):

- R.N. Noyce (Fairchild: future founder of Intel),

- Si monolithic circuit; several devices; resistors, capacitors, PN junction-insulation: Lehovec patent: Prague Electric Company,

- G. Moore → broadcast areas;

– 1958: J. Hoerni (Fairchild): diffused transistors (base and emitter diffused in the collector):

- passivation of junctions by oxide layer,

- lithographing and etching,

- batch processing (on the same wafer) several chips (dies) → CI sold by Texas and Fairchild.

I.1.7. *Field-effect transistor*

– 1951: W. Schockley; JFET (junction field effect transistor):

- Pb: unstable surface (electric charge carrier trap);

– 1958: S. Teszner (France): first JFET manufactured, thanks to the planar process (replaced mesas – trays – and passivation: S_iO_2);

– 1960: M. Atalla, D. Kahng (Bell Labs):

- first MOS (metal oxide-semiconductor), p-type: PMOS;

– 1962: first CMOS inverter; invention of CMOS technology by F. Wanlass, at Fairchild, technology distinguished by its low static consumption;

– 1962: S. Hofstein, F. Heiman (RCA) patent for the manufacture of MOS ICs (production of the first commercial field-effect transistors);

– 1963: CMOS: Complementary MOS: NMOS and PMOS;

– 1970: BiCMOS: CMOS compatible bipolar; for example, polycrystalline emitter of the NPN and manufactured at the same time as the polysilicon gate of the PMOS.

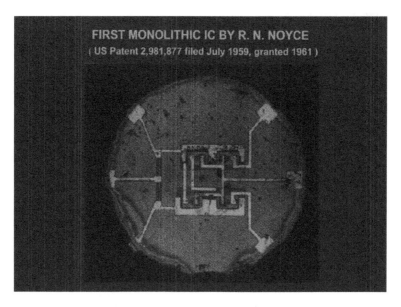

Figure I.1. *First integrated circuit (Noyce)*

I.1.7.1. *1951: Discrete transistors, by chip*

– 1960: Small-scale integration (100 devices);

– 1966: medium-scale integration (100 to 1,000 devices);

– 1969: large-scale integration (>1,000 devices);

– 1975: very large scale integration (VLSI) (>10,000 devices);

– 1986: ultra large scale integration (ULSI) (>1,000,000 devices);

– 2010: to 3D.

I.1.8. *Digital integrated circuits*

– 1961: J.L. Buie: TTL: transistor-coupled transistor logic (pacific semiconductor, ∈ TRW):

- TTL: transistor-transistor logic,

- for example, several emitters per transistor ==> stronger integration of devices;

– 1962: Motorola → ECL: emitter coupled logic;

– 1965: founding of Intel (Silicon Valley);

– 1967: ROM: Read only memory:

- PROM: Programmable ROM,

- EPROM: Erasable PROM;

– 1968: launch of the ARPANET network, the ancestor of the INTERNET;

– invention of the Windows-Mouse Environment (XEROS, ceded to Microsoft);

– 1969: M. Hoff (Intel) → microprocessor;

– UNIX/operating system;

– 1970: Bipolar RAM (recording ~1,000 bits (binary digit) of info);

– 1970: W.S. Boyle, G.E. Smith (Bell labs) → Memory and registers with 64000-bit RAM: 1977:

- application: cameras, image processing, telecom;

– 1971: Microprocessor 4 bits:

- creation of the first microprocessor: the Intel 4004 (clock rate of 740 kHz: 2,300 transistors);

– 1971: ion implantation: I^2:

- on-chip interconnect lengths:

- 1961: 25 mm,

- 1975: 2 mm,

- 1990: 1 mm,

- CAD: Computer-aided design,

- manufacturing process simulator:

  - SUPREM (Standford) 1978: first 1D simulator,

  - 2D BICEPS (finite differences),

  - TITAN (CNET Grenoble 3D finite elements: process),

  - ATHENA SILVACO;

– for the "device":

- PISCES/MEDICI,

- STORM (France–Italy: process and device; abandoned: because of bureaucracy),

  - ATLAS SILVACO,

  - SENTAURUS SYNOPSIS;

– electric circuit simulator (Kirchhoff's laws):

- SPICE: Simulation program with integrated circuits emphasis (L.W. Nagel, A. Vladimirescu),

  - ELDO: very suitable for CMOS; electrical simulator RF:

    - layout/schematic: CADENCE + ADS (advanced design system, by Hewlett Packard);

– 1971: M.J. Cochran and G. Bonne (Texas); patent for an on-chip "microcomputer" (however Intel's 8048 was the first available on the market):

- CCD (MOS): charge-coupled device ("multi-gate" MOS);

– 1972: Micro. 8 bits;

– 1972: K. Hart, A. Slob (Phillips, the Netherlands) and H.H. Berger, S.K. Wiedman (IBM, Germany) $\rightarrow$ I^{2}L (integrated injection logic (multi-collector transistors $\rightarrow$ very high-density bipolar chips)):

- CMOS: initial use in watches (Japan and Switzerland);

– 1973: 16,000 bits (MOS);

– 1973: creation of C programming language, closely related to the UNIX operating system (1969: K. Thompson and D. Ritchie);

– 1974: B. Wildar (Fairchild Semiconductor):

- first operational amplifier; the mA 709,

- most analog circuits had bipolar circuits (see outgoing speed), but MOSs have been used since the late 1970s,

- for example:

  - analogue multipliers (e.g. Gilbert),

  - A/D and analog digital converter,

  - phase-locked loop (PLL) and its VCO (voltage-controlled oscillator);

– 1976: launch of the CRAY 1 supercomputer (peak power 100 MFlops);

– 1977: first Apple computer;

– 1977: Micro. 16 bits, 32 bits then 64 bits;

– 1978: 64,000 bits;

– 1982: 288,000 bits;

– 1988: $>10^6$ bits;

– 1990: CMOS supremacy:

- MOS are widely used in RAM (random access memories): random access (write and read).

I.1.9. *Manufacturing technologies*

– 1960: epitaxy (epi: upon, taxis: arrangement < Greek);

– 1967: electron beam masking;

– 1971: ion implantation: I^2;

– 1981: IBM starts commercializing personal computers;

– 1983: creation of the C++ programming language;

– 1984: launch of Apple's Macintosh, the first commercial success of a computer with a "mouse-window" environment (XEROS technology);

– 1986: launch of the Windows 1.1 operating system; by Microsoft;

– 1989: creation of the World Wide Web and the HTML language (Hyper Text Markup Language) at the European Centre for Nuclear Research, CERN;

– 1994: Intel launches Pentium, a microprocessor containing more than 5 million transistors:

- novelties: data bus expands to 64 bits for memory access and processor capacity to be able to process two instructions per clock cycle and two levels of cache memory in order to accelerate the processing of instructions at the processor level;

– 1998: launch of Intel's Pentium II and AMD's K6-2;

– 2001: debut of the Intel Pentium III. This processor increased the frequency of PCs to 866 MHz;

– 2003: start of Intel's Pentium IV. Launched at 1 GHz, it reached up to 3.8 GHz;

– 2003: the number of transistors on a PC chip reaches billions. A PC microprocessor can deliver up to 6.4 GFlops (6.4 billion floating-point operations per second).

a) First μ processor: Intel 4004: 1971   b) INTEL P4: 2000

Figure I.2. *The supercomputer: IBM's Doe Blue Gene installed at Lawrence Livermore National Laboratory (USA); currently, the fastest: Fujistu/Riken (442 petaflops: 2022)*

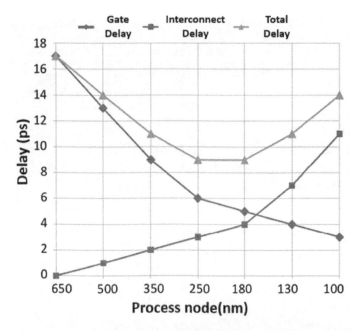

Figure I.3. *Delay time according to process node*

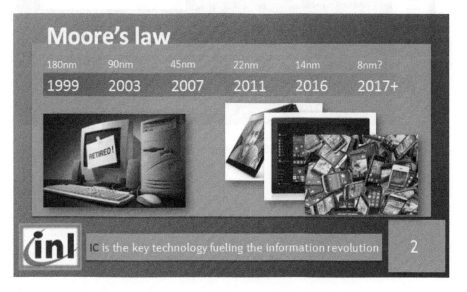

Figure I.4. *Moore's law (1975, the number of transistors of microprocessors (and no longer simple integrated circuits – case of the 1st conjecture of 1965) on a silicon chip doubles every two years)*

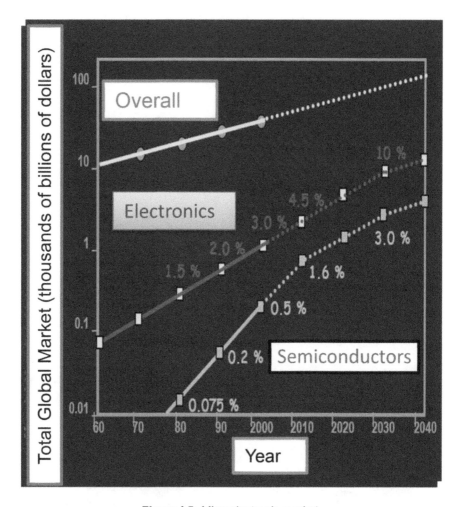

Figure I.5. *Microelectronic market*

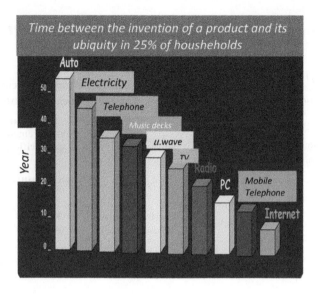

Figure I.6. *Penetration of microelectronics in the consumer market*

I.2. Computer-aided design

THE PLACE OF COMPUTER-AIDED DESIGN (CAD)

Ideal microelectronic CAD

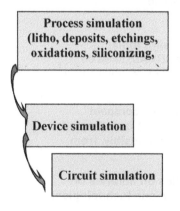

> Process simulation (litho, deposits, etchings, oxidations, siliconizing,

> Device simulation

> Circuit simulation

➢ **It is simulated**
➢ **It is manufactured**
➢ **It is measured**
... and it works on the first try!

Figure. I.7. *CAD*

The position of Computer Aided Design (CAD)

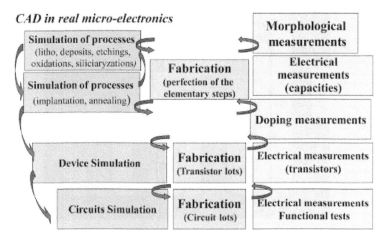

Figure I.8. *CAD: computer-aided design*

Computer Aided Design (CAD)

Actual microelectronic CAD:

> ➢ **Always at least one generation of gap between**
>> ➢ Innovative material: e.g. new dielectrics
>> ➢ Innovative device (transistor)
>> ➢ Innovative circuit
>> ➢ Innovative service
>> ➢ *The same goes for design tools!*
>> ➢ Technological simulation (TCAD) relies on physical models that need to be continuously refined and calibrated.

> ➢ **Circuit simulation is based on "compact" models whose parameters must be extracted from experimental data (and, possibly, from device simulation).**

> ➢ **Functional simulation relies on libraries of cells to be regenerated from one technological generation to the next.**

Figure I.9. *Microelectronic CAD*

Technology Computer-Aided Design (TCAD)

Actual microelectronic TCAD

100% predictive technology simulation does not exist!

> ➤ Physical mechanisms are becoming increasingly **complex.**
> ➤ Technological developments are **rapid** → modeling is difficult to follow.
> ➤ Calibrating a model always needs **experimental data.**
> ➤ Despite a very rigorous control of technological parameters in the silicon industry, their **fluctuations** are never insignificant.

Commercial simulators are more useful in industry than in research.

They are tools that are:
- powerful for interpolating between well-known situations;
- uncertain for extrapolating and exploring new concepts.

Figure I.10. *Technology computer-aided design*

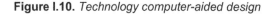

"TCAD" models vs "CAD" models

❑ TCAD models (numerical models).
❑ To take into account the technological parameters (T for technology).
❑ To "see" the inside of layers (semiconductors and dielectrics) → these are usually 2D or 3D models:
    ❑ **Physical** models based on partial differential equations;
    ❑ **Empirical** models based on more or less simple mathematical functions (polynomials, spline function, etc.).
❑ Analytical CAD (compact models) models.

Figure I.11. *Hierarchical templates*

Error due to the model itself = simplified representation of reality:

– cases where analytical solutions are sought.

Error sources:

– simplifying hypotheses to find exact solutions ➔ difficult to deduce the error induced on results;

– when looking for numerical solutions;

– meshes: in time, in space, etc. (real space and reciprocal space – velocities or wave vectors);

– finished number of iterations on:

- nonlinear problems,

- very large linear systems,

- coupled equation systems;

– numerical analysis provides some tools for estimating these errors.

I.3. Manufacturing: technological processes, diffusions: brief reminders

Here, we present a reminder of the manufacturing process of integrated electronic devices; in this case, we insist on dopant diffusions, which are crucial steps. Therefore, we will focus on the diffusion equation and its use in the context of modeling the manufacturing of microelectronic devices.

I.3.1. *Simulators for technological procedures*

Diffusion processes are surely the most crucial steps when manufacturing devices for integrated circuits. They are underpinned by the basic equation: the equation of diffusion, which is also that of heat.

Physicists and technologists were brought closer together in the 1970s, until it became unavoidable to confuse the two-dimensional problems – lateral diffusions, accelerated diffusion by oxidation, or by the presence of other dopants.

We should also bear in mind that the physics of point defects, metallic for example, in the context of semiconductors is far from being well taken into account in commercial simulators; this problem could be aggravated with regard to miniaturization.

Commands	Parameters		
Specification of models	Technological	Physical	Numerical
Description of substrate	Thickness	Anisotropy coeff.	Mesh size
Definition of masks	Dimension	Reaction coeff.	Time step
Deposit	Duration	Diffusivity	Number of iterations
Etching	Dose	–	Convergence criteria
Ionic implanation	Energy		Relaxation parameter
Annealing	Gas pressure		
Electrical contacts	Temperature		
Saving the results	–		

Table I.1. *Process modeling*

Since the 1980s, ion implantation has replaced the initial diffusions; the average depth of the implanted dopants is well controlled, and subsequent thermal annealing (rapid thermal annealing (RTA)) makes adjusting post-implantation diffusions (drive-in) possible.

The diffusion equation, for dopants in this case, is one of the fundamental equations of physics; it involves a famous operator: the Laplacian!

I.3.2. *Main stages of device manufacturing*

The main stages of device manufacturing are as follows:

a) ion implantation: I^2;

b) drive-in;

c) thermal oxidation;

d) epitaxial growth;

e) etching;

f) deposition, oxide.

1.4. PN junction

The definitions are as follows:

– PN junction: inside the same crystal, the semiconductor goes from type "P" to type "N".

– Homojunction: the "P" type semiconductor is made of the same material (Si, Ge, GaAs) as the "N" type semiconductor. Otherwise, it is referred to as heterojunction.

– Metallurgical junction: the plane where the semiconductor changes type.

– In the one-dimensional model, the distribution of impurities is studied only along an Ox axis:

- the impurity profile is the difference between the density of the acceptor atoms and the density of the donor atoms ($N_A - N_D$);

- for simplicity, it will be assumed that on the "P" side, the excess of the density of the acceptor atoms is equal to N_A (cm^{-3}) and, on the "N" side, the excess of the density of the donor atoms is equal to N_D (cm^{-3});

- step junction: the transition from the "P" region to the "N" region takes place over an "infinitely" thin thickness;

- linearly graded junction: the transition from the "P" region to the "N" region takes place according to a linear law.

– In reality, the doping profile is fairly well represented by the erfc(x) function: this is the real junction.

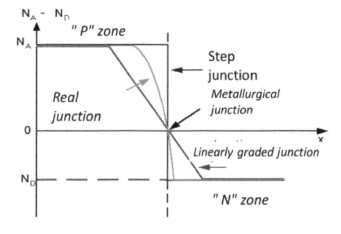

Figure I.12. *Different junction profiles*

The PN diode has a "P" semiconductor region, a PN junction and an "N" semiconductor region.

In practice, modern PN diodes are highly asymmetrical: doping of one of the zones is much more significant than the other (to improve the injection of carriers).

If the P side is much more doped than the N side, it is a P^+N diode.

If the N side is much more doped than the P side, it is a PN^+ diode:

– With reference to the vacuum diode, the "P" part of the diode is sometimes called the anode.

– The PN diode has the specific feature of allowing the current to flow in only one direction.

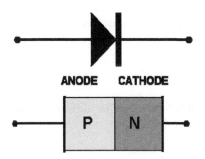

Figure I.13. *The symetrized K of the symbolic diagram of the diode; cathode: region N*

The unpolarized PN junction:

- Consider a PN step junction (constant P doping = N_A, constant N doping = N_D) at thermodynamic equilibrium (ideal junction).

Let us imagine that the semiconductor "P" is separated from the semiconductor "N".

In semiconductor "P", the FERMI level (maximum electrochemical energy of the holes) is located at a distance above the maximum of the valence band, such that:

$$\partial E_p = E_{FP} - E_V = kT.Log\left(\frac{N_V}{N_A}\right)$$ [I.1]

In semiconductor "N", the FERMI level is below the CB minimum at a distance such that:

$$\partial E_n = E_C - E_{FN} = kT.Log\left(\frac{N_C}{N_D}\right) \tag{I.2}$$

If the two semiconductors are part of the same crystal lattice and at thermodynamic equilibrium (no polarization), the FERMI levels align.

$$E_G = qV_b + \partial E_n + \partial E_p \quad \text{(G: Gap)} \tag{I.3}$$

There is a distortion of the energy bands. The difference between the minimum of the conduction band (CB) on the P side and the minimum of the CB on the N side corresponds to the variation of the potential energy of the conduction electron. This results in the appearance of a potential barrier:

$$qV_b = E_G - k_B T \ Log\frac{N_C N_V}{N_A N_D} \tag{I.4}$$

Here, V_b is the barrier potential or diffusion potential (V_{bi} is the built-in potential, diffusion potential); therefore:

$$qV_b = k_B T \ Log\frac{N_A N_D}{n_i^2} \tag{I.5}$$

There is a variation of the V_{bi} potential (bi: built in potential) by crossing a PN junction, even if the external polarization is zero.

With identical N_A and N_D dopings, plus E_G ↗ plus V_{bi} ↗.

Potential barrier heights are high in high bandgap semiconductors (e.g. SiC, GaN).

For fixed E_G, plus the product $N_A N_D$ ↗ plus V_{bi} ↗:

The greater the dopings of the P and N parts, the greater the barrier potential of the junction.

When the operating temperature of the junction increases, E_G hardly varies and the height of the potential barrier decreases.

I.4.1. *Forward-biased PN junction*

Direct polarization: bring the "P" part of the junction to a positive potential with respect to the "N" part.

We can note that V_j is the potential difference created by the external source at the junction. For a forward bias, this quantity V_j (j: junction) is positive.

By applying a direct polarization, the height of the potential barrier that existed when the junction was not polarized is reduced.

$$E'_M = E_M \sqrt{1 - \frac{V_j}{V_b}} \quad (V/m) \tag{I.6}$$

By applying a direct polarization, the electric field of retention of diffusion is reduced; consequently, the more a junction is forward-biased, the greater the diffusion of the holes from the "P" region toward the "N" region and the diffusion of electrons from the "N" region toward the "P" region. The direct current appears.

I.4.2. *Reverse-biased PN junction*

The "P" part of the junction is at a negative potential with respect to the "N" part.

V_j, the potential difference created by the external source at the junction, is negative.

– The potential of the "N" part is the same; in the energy diagram above, the FERMI level on the "N" side does not move.

– Energies increase upwards; for a decrease in the potential, $-qV$ is positive, so the energy level "rises".

– When the negative potential difference V_j is applied to the junction, the FERMI level on the "P" side rises by quantity qV_j.

– The height of the potential barrier, which was qV_b without polarization, is now greater and it is equal to:

$$qV_b = qV_{bi} - qV_j \tag{I.7}$$

By applying a reverse polarization, the height of the potential barrier that existed when the junction was not polarized is increased.

The more the reverse bias V_j (increasingly negative) is increased, the more the non-neutral SCZ (space charge zone, deserted zone), on either side of the metallurgical junction, increases, and therefore the more the capacity of the junction decreases.

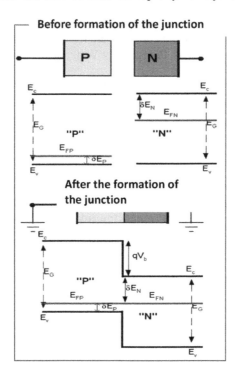

Figure I.14. *Junction without bias*

Generally:

$$I_D = I_S \left(\exp \frac{qV_j}{k_B T} - 1 \right)$$

[I.8]

but since V_j is negative, $\exp(qV_j/k_B T)$ tends toward 0, and the reverse current is then: $I_i = -Is$.

NOTE.– The reverse current: the minority carriers "see" the direct junction.

The reverse current flowing through a reverse-biased junction is very low (<mA), which is independent of the applied voltage and varies greatly with temperature (twice every 8°C).

Therefore, it comes from minority carriers (p in the N region, n in the P region). These minority carriers, which "see" the direct junction, have a concentration of $\approx 10^{10}$ cm^{-3} at ambient temperature.

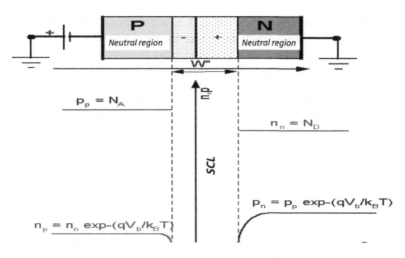

Figure I.15. *Reverse junction; carrier profiles (space charge zone (SCZ))*

What happens at the PN junction when it is reverse biased?

Figure I.18 shows that at the instant when the circuit is closed, some of the free electrons leave the N zone of the crystal and move toward the positive pole of the power supply battery.

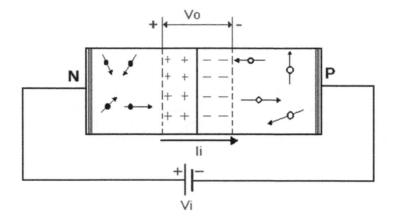

Figure I.16. *Reverse diode; carrier displacement*

At the same time, a certain amount of electrons emitted by the negative of the battery join the P zone of the crystal, where they will make part of the holes disappear.

Now, if we accept that, in the P zone, there are no free electrons which can join the N zone to replace those which are pushed back toward the positive of the battery and that, in the N zone, there are no holes that can propagate as far as the P zone in order to replace those which have disappeared, we could conclude that the movement of the charges circulating from the crystal to the battery and from the battery to the crystal has ceased. Indeed, the number of free electrons present in the N zone of the crystal is certainly very large, but not unlimited; the same is true for the holes present in the P crystal.

In reality, the displacement of charges and consequently, the current produced by the power supply (battery), ceases even before the N zone has been freed from its electrons and the P zone from its holes.

To explain this phenomenon, it should be noted that the potential barrier strengthens rapidly with the decrease in free electrons and holes in the respective zones and its amplitude increases from Vo to Vo'.

The new potential difference Vo' can thus cancel the effect of the external voltage Vi, before all the electrons of the zone N are pushed back toward the positive of the battery and all the holes of the zone P have disappeared.

The voltage Vi applied to the terminals of the diode is called reverse voltage. If we consider the above, current flowing in the diode (at the terminals of which a reverse voltage has been applied) should cancel out rapidly. In reality, the current is not completely canceled out because of the presence of the minority carriers, that is because of the presence of holes in the zone N (with p< >n). It is said that the minority carriers "see" the junction directly.

A certain number of minority carriers always manage to pass through the junction, thus causing a partial replacement of free electrons in the N zone and of the holes in the P zone. Therefore, the presence of a very low current ($\sim 10^{-15}$ A at ambient temperature) is observed, flowing from the N end to the P end of the crystal. This current is called reverse saturation current (Is).

REVERSE N$^+$P STEP JUNCTION

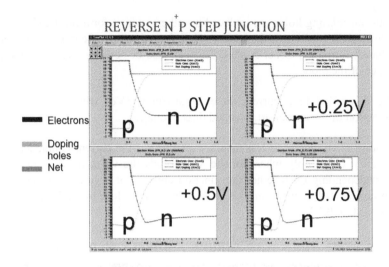

Electrons

Doping
holes
Net

Figure I.17. *n and p carrier density*

Figure I.17 shows reverse-biased N$^+$P step junctions, at 0, 0.25, 0.5, and 0.75 Volts (SILVACO simulations).

Let us go back to the PN junction when it is forward biased.

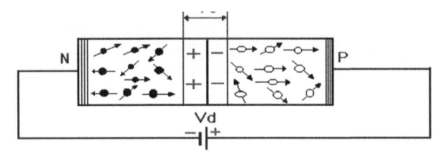

Figure I.18. *Direct diode; carrier displacement*

When the circuit is closed, the electromotive force of the battery sets free electrons of the N zone and holes of the P zone in motion, which converge toward the junction, inside which electrons fall into holes, which causes both to disappear. However, free electrons that fall into holes are continually replaced by others, coming from the negative pole of the power source.

Thus, all missing holes are replaced by others, which form on the side of the P zone, toward the positive of the battery. Therefore, the flow of charges is perpetually reproduced, forming a direct current. This can also be seen by measuring the direct resistance of the diode.

The Id is called direct current, the external voltage (Vd), which is at the origin of the formation of the current Id, is called direct voltage.

As long as voltage V_d is less than or equal to V_{bi}, the current is practically zero. This current exists only when voltage V_d exceeds the value of V_{bi}. This value is different depending on whether the junction consists of a germanium crystal or a silicon crystal: for germanium, this value is normally close to 0.25 V, whereas for silicon it is 0.6–0.7 V.

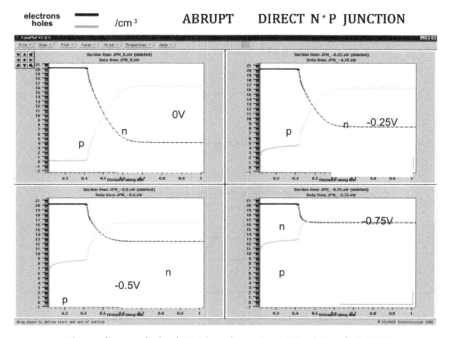

Abrupt direct polarized N$^+$P junction at 0, - 0.25, - 0.5 and- 0.75 V

Figure I.19. *n and p carrier density (SILVACO simulations)*

A PN junction allows a current to pass when the latter flows through the semiconductor in the direction from the P-doped crystal to the N-doped crystal. It opposes the circulation of a current in the opposite direction.

1

Bipolar Junction Transistor

1.1. Introduction

The bipolar junction consists of three zones doped with N-type impurities, P-type impurities and then N-type impurities to form an NPN or P-type bipolar transistor, N-type impurities and then P-type impurities to form a PNP-type transistor.

Three electrodes that are the contacts of base (B), emitter (E) and collector (C) are placed on these three zones.

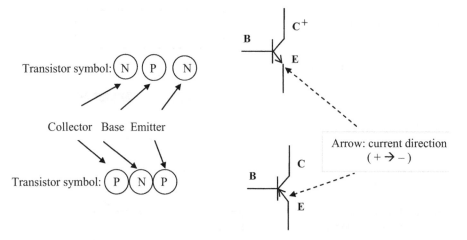

Figure 1.1. *Symbols of bipolar junction transistors*

For a color version of all figures in this chapter, see http://www.iste.co.uk/gontrand/analog. zip.

The properties of the bipolar junction transistor derive from those of the PN junction, as it essentially consists of two junctions: base-emitter and base-collector.

Convention on the deflection of currents and voltages:

– (e.g. symbol of the NPN here): this is the direction of the current;

– from + to - in the branch considered;

– (mnemonic medium; "he who can do the most, can do the least").

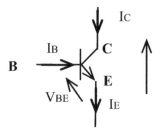

Figure 1.2. *NPN: currents deflection*

1.1.1. *A schematic technological embodiment of an integrated bipolar junction transistor*

Preparation of a substrate (substrate, bulk) P

Oxidation: photo-etching of the oxide implantation, then diffusion of the sole or buried layer of the collector; it is highly doped, therefore not very resistive.

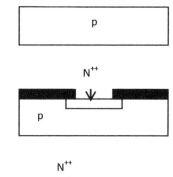

Oxide removal epitaxy (epi: above, taxis: order) of an N⁻ layer (collector), lightly doped

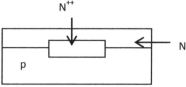

Oxidation: photo-etching of the oxide and then diffusion P of insulation walls (channel stops)

Oxidation: photo-etching of the oxide followed by base P$^+$ diffusion and recharging of the insulation walls

Oxidation: photo-etching of the oxide; N^{++} diffusion of the emitter and collector contact

Opening contacts:

Metallization: includes etching of contacts

Then encapsulation, BPSG protection:
Borophosphosilicate glass (ductile glass – see mechanical stresses, relaxed thanks to boron)

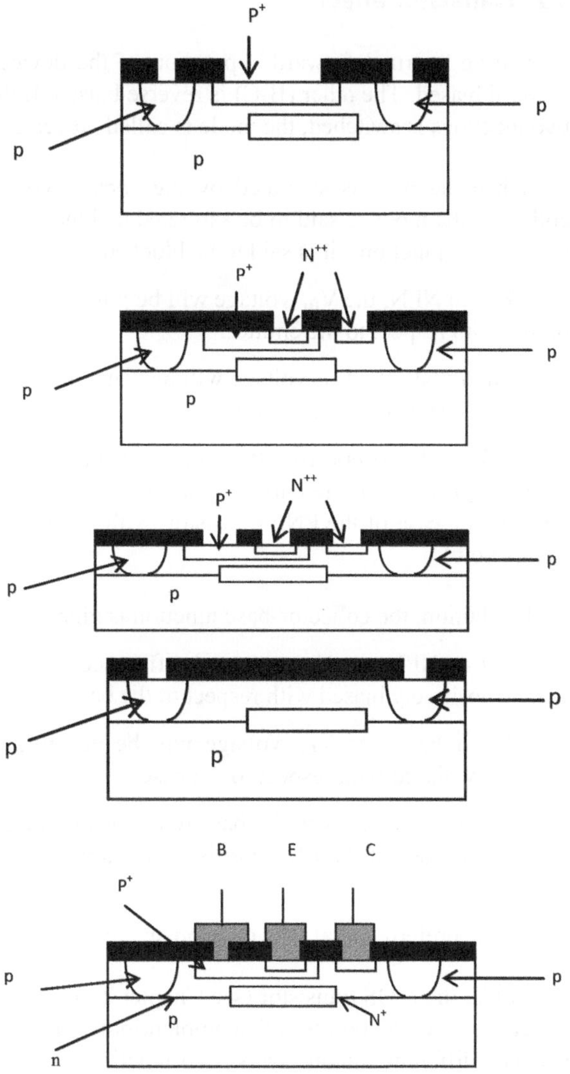

Figure 1.3. *Fabrication steps of a bipolar junction transistor*

1.2. Transistor effect

During "normal forward" operation of the device, the base-emitter junction is forward biased. The other (B-C) is reverse biased. If the polarization direction of the two junctions is switched, the mode is called "reverse normal".

When the base is saturated by the carriers coming from the emitter and the collector, the mode is said to be saturated and the two junctions are forward. For the two reverse junctions, it is said to be blocked.

– For an NPN, the V_{BE} voltage will be positive, that is, the base will be positively biased with respect to the emitter.

– For a PNP, the V_{BE} voltage will be negative, that is, the base will be negatively biased with respect to the emitter:

- flow of carriers from the emitter to the base (of electrons in the case of the NPN transistor, with the current going in the opposite direction to the flow, and of holes in the case of the PNP transistor, with the current going in the same direction as this flow).

In addition, the collector-base junction is highly reverse biased (a few volts).

– For an NPN, the V_{CB} potential difference will be positive, that is, the collector will be positively biased with respect to the base.

– For a PNP, the V_{CB} voltage will be negative, that is, the collector will be negatively biased with respect to the base:

- minority carriers in the base (electrons in the case of the NPN transistor and holes in the case of the PNP transistor) are attracted (caught) by the potential of the collector;

- the preferred goal is to develop an amplifier.

Unlike the MOS transistor (see Chapter 2), the two types of carriers contribute to the current; conduction is called ambient-polar. This current (often) flows vertically between different silicon layers: two junctions are formed on either side of a thin, moderately doped buried layer called the base, so that the non-polarized transistor does not conduct.

The very highly doped layer, of the type opposite the doping of the base and situated close to the surface, is called the emitter; it is formed by a counter-doping, sometimes implanted through a polysilicon deposit. The layer under the base, which is very lightly doped (therefore, very resistive), and of the same type as the emitter, is called the collector. The base and collector contacts must, therefore, be remote, which gives rise to the problem of access resistances.

Under "normal" conditions, voltages are applied to the contacts such that the base-emitter junction is conducting and the base-collector junction is blocking. However, if the base is sufficiently thin, the majority of electrons supplied by the emitter pass through the base-collector junction, and the current which passes from the emitter to the collector remains proportional, over several decades, to the current which passes from the emitter to the base, with a coefficient of proportionality called transistor gain, which may be much greater than 100: it is thus possible to amplify the signal represented by the emitter-base current.

The transistor phenomenon essentially comes from the small region of the base and the high doping of the emitter; the emitter current I_E which arrives at the base depends on the direct voltage applied V_{BE}; the holes in the case of the PNP arriving at this base are trapped by the high electric field created by the strongly negative voltage V_{CB} (in the case of PNP). A large part of these holes passes through the narrow base to penetrate into the collector and form I_C. Thus, this collector current depends relatively little on V_{CB}, provided that this polarization is sufficient.

1.2.1. *Flows and currents*

BIPOLAR TRANSISTOR - PRINCIPLE

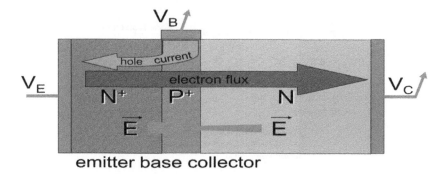

Figure 1.4. *NPN bipolar transistor: currents*

Bipolar junction transistor

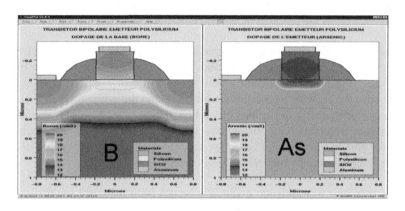

Figure 1.5. *SILVACO simulations (finite elements)*

1.2.2. *Compromises for bipolar junction transistor*

BIPOLAR COMPROMISES

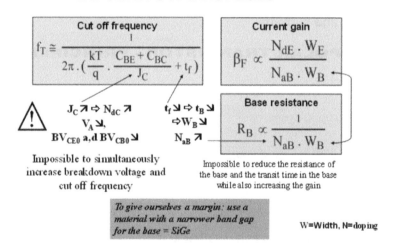

Figure 1.6. *Electrical parameters:*
compromises

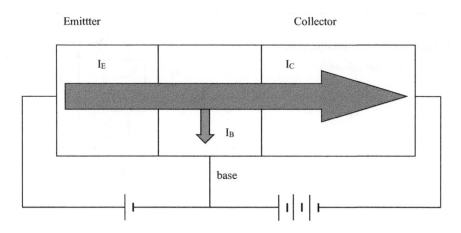

Figure 1.7. *Schematic example of the PNP transistor in forward normal mode (junctions: forward BE, reverse BC)*

1.2.3. *Configurations and associated current gains*

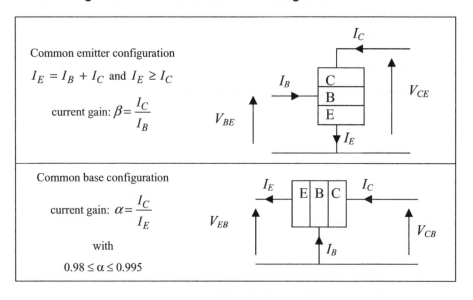

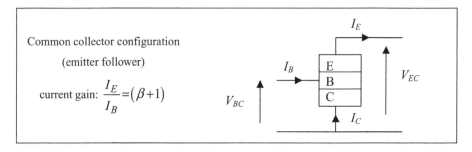

Figure 1.8. *Various configurations of the "bipolar"*

Note that $\beta = \dfrac{\alpha}{1-\alpha} \gg 1$

1.2.3.1. *Current gain control*

We can consider three determining factors to calculate this gain:

– injection efficiency: γ;

– transport factor: B_T;

– multiplication factor at the crossing of the base-collector junction (in reverse): M.

The current gain α is the product of these three factors:

$$\alpha = \gamma \; B_T \; M \tag{1.1}$$

Injection efficiency γ

This is the ratio between the current injected from the emitter to the base and the total emitter current. Good injection efficiency corresponds to γ close to unity. This is obtained in particular by virtue of a highly doped emitter (with respect to the base).

Transport factor B_T

This is the ratio of the current reaching the base-collector junction (reverse biased) to the current injected by the emitter. A good transport factor is B_T close to unity. This is obtained with a thin base (thickness much less than a micrometer).

Multiplication factor M

The fairly strong reverse polarization of the base-collector junction may allow carriers to acquire a sufficiently high energy to cause the impact ionization phenomenon. This leads to the creation of new electron–hole pairs and therefore an additional current at the collector input.

This is a parasitic phenomenon, which is inappropriate because it introduces a link between the current gain and the bias of the collector; this has nothing to do with the transistor effect. We need to be careful as there is no longer any control of Ic by Ib (if M >1).

1.3. Bipolar junction transistor: some calculations

Minority carriers obey the (famous) diffusion law for a PNP:

$$\frac{\partial p}{\partial t} = D_p \frac{\partial^2 p}{\partial x^2} - \frac{p - p_0}{\tau_p} \qquad [1.2]$$

We are interested in the concentration of electrons injected from the base into the emitter:

$$\frac{\partial^2 N_E}{\partial x^2} - \frac{N_E - n_{pE}}{L_{nE}^2} = 0 \qquad [1.3]$$

where:

– $N_E(x)$: concentration of electrons in the emitter (at ambient temperature, donors are all ionized);

– n_{pE}: concentration of minority carriers (electrons), in the absence of injection, in the emitter;

– L_{nE}: diffusion length of minority cess (electrons) in the emitter.

The limit condition for $N_E(x)$ at the emitter-base junction is as follows:

$$N_E(-W/2) = n_{pE}.e^{V_{EB}/U_T} \qquad [1.4]$$

$U_T = kT/q$: thermal voltage: ~25 mV at ambient temperature (300 K).

The solution of this diffusion equation is of the form:

$$N_E(x) - n_{pE} = A.e^{x/L_{nE}} + B.e^{-x/L_{nE}} \qquad [1.5]$$

For Cl:

$$N_E(x) = \frac{n_{pE}\left(e^{\frac{V_{EB}}{U_T}} - 1\right).sh(\frac{l_E + x)}{L_{nE}})}{sh(\frac{L_E - W/2)}{L_{nE}})} + n_{pE} \tag{1.6}$$

The electron current at the emitter is given as:

$$I_{nE} = Aq\, D_{nE}\left(\frac{d\,N_E}{dx}\right)_{x= -W/2} \tag{1.7}$$

$$= Aq\, D_{nE}\frac{n_{pE}}{L_{nE}}\left(e^{\frac{V_{EB}}{U_T}} - 1\right)\coth\frac{l_E - W/2}{L_{nE}} \tag{1.8}$$

Usually:

$$l_E - W/2 \text{ (emitter width, W; base width)} \gg L_{nE} \tag{1.9}$$

$$I_{nE} = Aq\, D_{nE}\frac{n_{pE}}{L_{nE}}\left(e^{\frac{V_{EB}"}{U_T}} - 1\right) \tag{1.10}$$

– Distribution of holes injected by the emitter and the collector into the base.

Let $p_n(x)$ be the concentration of holes in the base; it obeys the following equation of continuity:

$$\frac{\partial^2\, P_B}{\partial x^2} - \frac{P_B - p_{nB}}{L_{pB}^2} = 0 \tag{1.11}$$

The solution is given as:

$$P_B(x) - n_{pB} = A.e^{x/L_{pB}} + B.e^{-x/L_{Pb}} \tag{1.12}$$

The constants A and B are deduced from conditions at the following limits:

$$P_B(-W/2) = p_{nB}.e^{\frac{V_{EB}}{U_T}} \tag{1.13}$$

and

$$P_B(+W/2) = p_{nB}\, e^{\frac{V_{CB}}{U_T}} \tag{1.14}$$

Hence, the solution is given as:

$$P_B(x) - p_{nB} = \frac{p_{nB}(e^{\frac{V_{EB}}{U_T}} - 1)\, sh\frac{\frac{W}{2} - x}{L_{pB}} + p_{nB}(e^{\frac{V_{CB}}{U_T}} - 1)\, sh\frac{W/2 + x}{L_{pB}}}{sh\frac{W}{2}/L_{pB}}$$ [1.15]

This equation can be considered as the sum of two separate injections: one coming from the emitter, with the collector being short-circuited ($V_{CE} = 0$), and the other coming from the collector, with $V_{BE} = 0$.

The operation of a transistor in the most general case where $V_{CE} > 0$ can therefore be considered as the result of an injection at the emitter, with the collector being short-circuited, and injection at the collector, with the emitter being short-circuited.

The reverse hole current at the emitter is given by:

$$I_{pE} = -Aq\, D_{pB}\left(\frac{d\, p_B}{dx}\right)_{x = -\frac{W}{2}}$$ [1.16]

We have:

$$I_{pE} = Aq\, D_{pB}\frac{p_{nB}}{L_{pB}}\coth\left(\frac{W}{L_{pB}}\right)\left(e^{\frac{V_{EB}}{U_T}} - 1\right) - Aq\, D_{pB}\frac{p_{nB}}{L_{pB}sh\frac{W}{L_{pB}}}\left(e^{\frac{V_{CB}}{U_T}} - 1\right)$$ [1.17]

The total emitter current is:

$$I_E =$$
$$\left(Aq\, D_{pB}\frac{p_{nB}}{L_{pB}}\coth\frac{W}{L_{pB}} + Aq\, D_{nE}\frac{n_{pE}}{L_{nE}}\right)\left(e^{\frac{V_{EB}}{U_T}} - 1\right) -$$
$$Aq\, D_{pB}\frac{p_{nB}}{L_{pB}sh\, W/L_{pB}}\right)\left(e^{\frac{V_{CB}'}{U_T}} - 1\right)$$ [1.18]

The calculations are similar for the collector.

We start from:

$$\frac{\partial^2 N_C}{\partial x^2} - \frac{N_C - n_{pC}}{L_{nC}^2} = 0$$ [1.19]

Finally:

$$
I_E =
$$
$$
\left(Aq\, D_{pB}\, \frac{p_{nB}}{L_{pB}}\, \coth\left(\frac{W}{L_{pB}}\right) + Aq\, D_{nE}\, \frac{n_{pE}}{L_{nE}} \right)\left(e^{\frac{V_{EB'}}{U_T}} - 1 \right) -
$$
$$
Aq\, D_{pB}\, \frac{p_{nB}}{L_{pB}\mathrm{sh}\, W/L_{pB}} \right)\left(e^{\frac{V_{CB}}{U_T}} - 1 \right) \tag{1.20}
$$

$$
- I_C =
$$
$$
\left(Aq\, D_{pB}\, \frac{p_{nB}}{L_{pB}}\, \coth\left(\frac{W}{L_{pB}}\right) + Aq\, D_{nC}\, \frac{n_{pC}}{L_{nC}} \right)\left(e^{\frac{V_{CB}}{U_T}} - 1 \right) -
$$
$$
Aq\, D_{pB}\, \frac{p_{nB}}{L_{pB}\mathrm{sh}\, W/L_{pB}} \right)\left(e^{\frac{V_{EB'}}{U_T}} - 1 \right) \tag{1.21}
$$

Let us consider the regime in direct normal operation.

For an NPN, we have negative V_{CE} and positive V_{BE} (forward base-emitter junction and reverse base-collector junction).

In addition:

$$
|V_{CE}| \gg 0
$$

It is then possible to ignore exponentials containing V_{CE} in front of the unit. Then:

$$
I_E =
$$
$$
\left(Aq\, D_{pB}\, \frac{p_{nB}}{L_{pB}}\, \coth\left(\frac{W}{L_{pB}}\right) + Aq\, D_{nE}\, \frac{n_{pE}}{L_{nE}} \right)\left(e^{\frac{V_{EB''}}{U_T}} - 1 \right) +
$$
$$
Aq\, D_{pB}\, \frac{p_{nB}}{L_{pB}\mathrm{sh}\, W/L_{pB}} \right). \tag{1.22}
$$

$$
- I_C =
$$
$$
\left(Aq\, D_{pB}\, \frac{p_{nB}}{L_{pB}}\, \coth\left(\frac{W}{L_{pB}}\right) + Aq\, D_{nC}\, \frac{n_{pC}}{L_{nC}} + Aq\, D_{pB}\, \frac{p_{nB}}{L_{pB}\mathrm{sh}\, W/L_{pB}} \right)\left(e^{\frac{V_{EB'}}{U_T}} - 1 \right) \tag{1.23}
$$

– Transistor DC gain:

This gain is the ratio of the output collector current to the emitter current generating it.

$$\alpha_0 = \frac{d(-I_c)}{d\, I_I}$$

We have:

$$\alpha_0 = \frac{1}{ch\frac{W}{L_{pB}}(1+\frac{D_{nE}\, n_{pE}\, L_{pB}}{D_{pE}\, p_{nB}\, L_{nE}}th\frac{W}{L_{pB}})} \qquad [1.24]$$

Often $W \ll L_{pB}$

$$\alpha_0 = \frac{1}{ch\frac{W}{L_{pB}}(1+\frac{D_{nE}\, n_{pE}\, W}{D_{pE}\, p_{nB}\, L_{nE}})} \qquad [1.25]$$

However:

$$\alpha_0 = \beta_0^*\cdot \gamma_0 \qquad [1.26]$$

– Transport factor:

$$\beta_0^* = \frac{1}{ch\frac{W}{L_{pB}}} \qquad [1.27]$$

If $W \ll L_{pB}$:

$$\beta_0^* = 1 - \frac{1}{2}\left(\frac{W}{L_{pB}}\right)^2 \qquad [1.28]$$

– Emitter yield: ratio of the current of holes injected by the emitter into the base to the total emitter current:

$$\gamma_0 = \frac{1}{1+\frac{D_{nE}\, n_{pE}\, W}{D_{pE}\, p_{nB}\, L_{pB}}} \quad \left(=\frac{I_{pE}}{I_{pE}+I_{nE}}\right) \qquad [1.29]$$

By ignoring $e^{\frac{V_{CB}}{U_T}}$ in front of 1 and 1 in front of $e^{\frac{V_{EB}}{U_T}}$, and if: $\frac{W}{L_{pB}} \ll 1$, we get:

$$I_{pE} = Aq\, D_{pB}\frac{p_{nB}}{W}e^{\frac{V_{EBI}}{U_T}} \qquad [1.30]$$

Similarly:

$$I_{nE} = Aq\, D_{nE}\, \frac{n_{pE}}{W}\, e^{\frac{V_{EB'}}{U_T}} \qquad [1.31]$$

NOTE.– We can highlight a remarkable result:

Based on resistivities:

$$\rho_E = \frac{1}{qP_{pE}\mu_{pE}} = \frac{n_{pE}}{q\mu_{pE}n_i^2} \text{ and } \rho_B = \frac{1}{qn_{nE}\mu_{nB}} = \frac{P_{nB}}{q\mu_{nB}n_i^2}, \qquad [1.32]$$

And considering:

$$n_i^2 = n_n p_n = p_p n_p \qquad [1.33]$$

We have:

$$\gamma_0 = \frac{1}{1 + \frac{\rho_E}{\rho_B}\frac{W}{L_{nB}}} \qquad [1.34]$$

Note that g_0 is all the stronger when the emitter is highly doped and the base is weakly doped (but can be affected with regard to the noise of the device).

Then, we can write, in the normal direct mode, the hole density (PNP) in the base (ISE-TCAD Manuals, 2002):

$$P_B(x) = . \frac{P_{nB}e^{\frac{V_{EB}}{U_T}}\,sh\,\frac{\frac{W}{2}-x}{2} + P_{nB}(e^{\frac{V_{CB}}{U_T}})\,sh\,\frac{W/2+x}{2}}{sh\,W/L_{pB}} \qquad [1.35]$$

In normal direct mode:

$$P_B(x) = . \frac{P_{nB}\,(sh\,\frac{\frac{W}{2}-x}{L_{pB}})\,(e^{\frac{V_{EB}}{U_T}})}{sh\,W/L_{pB}} \qquad [1.36]$$

Note that if $W/2 - x$ is small compared to L_{pB}, a limited development can be made to the order of 1. The concentration of the holes is then in $-x$ (linearity). The corresponding diffusion current, proportional to the derivative of this concentration, will then be more important than if $L_{pB} \ll W$ (then, holes have a better ability to recombine in said base).

It is then possible to redo this type of calculation by adding a sinusoidal term (see "Bode"), which is small in relation to the corresponding continuous value. And so, we access transfer functions: gains, etc.

1.3.1. *Various modes of operation*

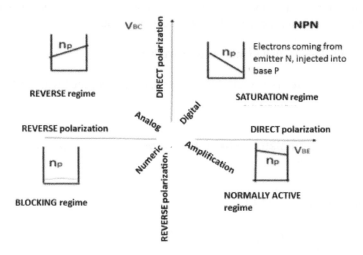

Figure 1.9. *The different modes of operation*

The study of a device, if considered microscopically, can be carried out using basic equations, drift-diffusion type equations, or master equations, such as the Boltzmann equation. But we must not forget that it is studying circuits, using these devices, that is the ultimate goal.

It is currently out of the question to use numerical approaches, typically differences or finite elements, for tens, thousands, or even more of the interconnected devices constituting circuits; the computation time would be prohibitive, as would the problems of divergence.

The idea, dear to industrialists, is to create so-called compact models, either from results derived from equations governing the microscopic world of these devices or from the actual manufacture of demonstrators, and/or from the output of electrical parameters by dedicated algorithms (e.g. via CADENCE ICCAP). A file is then obtained, describing a circuit typically comprising resistors, capacitors, inductors, current and "plated" voltage sources on the design, geometry, and appearance of the device and its layout.

These parameters, called SPICE (Simulation Program with Integrated Circuits Emphasis) parameters, then allow for the study circuits, therefore comprising several devices via the solution of Kirchoff-type equations (laws of nodes and the meshes). The study can then be done in static, alternative, transient, temperature or even noise.

Let us return to the bipolar junction transistor; it can be considered as a pair of interactive PN junctions, even if the latter are autonomous.

They are close enough to influence each other.

In practice, there are two versions of this model, associated with NPN and PNP transistors.

1.4. The NPN transistor; Ebers–Moll model (1954: Jewell James Ebers and John L. Moll)

$$I_F = I_{ES}\left(e^{\frac{q\,V_{BE}}{kT}} - 1\right)$$

$$I_R = I_{CS}(e^{\frac{q\,V_{BC}}{kT}} - 1)$$

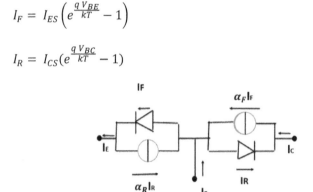

Figure 1.10. *Bipolar injection model (F: forward, R: reverse)*

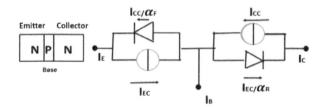

Figure 1.11. *Transport model (F: forward, R: reverse)*

In this case, which is better suited to computer simulation, I_{CC} and I_{EC} are current sources, which are written as follows:

$$I_{CC} = I_S(e^{\frac{q\,V_{BE}}{kT}} - 1) \tag{1.37}$$

$$I_{EC} = I_S(e^{\frac{q\,V_{EC}}{kT}} - 1) \tag{1.38}$$

The output currents are then given as:

$$I_C = I_{CC} - \frac{I_{EC}}{\alpha_R} \tag{1.39}$$

$$I_E = I_{EC} - \frac{I_{CC}}{\alpha_F} \tag{1.40}$$

$$I_B = -(I_C + I_E)$$

$$= (\frac{1}{\alpha_F} - 1)I_{CC} + (\frac{1}{\alpha_R} - 1)I_{EC} \tag{1.41}$$

Replacing the two currents refers to a simple source of current (I_{CT}) between the emitter and the collector:

$$I_{CT} = I_{CC} - I_{EC} = I_S\left(e^{\frac{q\,V_{BE}}{kT}} - e^{\frac{q\,V_{BC}}{kT}}\right) \tag{1.42}$$

The saturation currents of diodes are then changed.

These diode currents become:

$$\frac{I_{CC}}{\beta_F} = \frac{I_S}{\beta_F}\left(e^{\frac{q\,V_{BE}}{kT}} - 1\right) \tag{1.43}$$

$$\frac{I_{EC}}{\beta_R} = \frac{I_S}{\beta_R}\left(e^{\frac{q\,V_{BC}}{kT}} - 1\right) \tag{1.44}$$

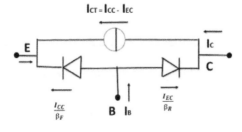

Figure 1.12. *"SPICE" model of bipolar NPN*

Output currents are then written as follows:

$$I_C = I_{CT} - \frac{I_{EC}}{\beta_R} \qquad\qquad [1.45]$$

$$I_E = -I_{CT} - \frac{I_{CC}}{\beta_F} \qquad\qquad [1.46]$$

$$I_B = \frac{I_{CC}}{\beta_F} + \frac{I_{EC}}{\beta_R} \qquad\qquad [1.47]$$

The model is very simple, essentially requiring only three parameters as follows:

a_F, a_R, Is.

It is accurate for a steady state regime or for a small range of polarizations.

1.4.1. *Gummel curves*

These are obtained by extrapolation of the $\log I_E$ as a function of V_{BE}, in the direct region, and $\log I_C$ as a function of V_{BC} in the inverse region.

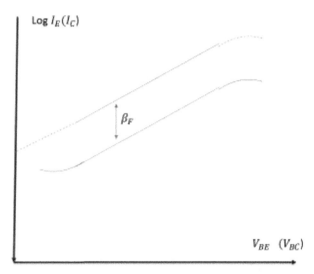

Figure 1.13. *Gummel curves*

1.4.2. *Consideration of second-order effects for the static model*

a) Parasitic elements for base, collector and emitter resistors:

r_E, r_C, r_B (between active zone and contact);

b) I_C dependence as a function of V_C (Early effect; V_A: base width modulation parameter);

c) a_F, current dependent a_R;

d) effect of high injection in the base and collector regions.

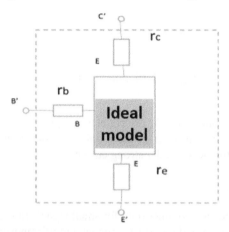

Figure 1.14. *A more complete model: access resistors*

– Access resistances (ohmic)

These access resistances connect the external "leads" to the internal model, and they improve (and complicate) the characterization of the model.

– Collector resistance r_C

It decreases the slope of characteristics $I_C(V_{CE})$ in the region of quasi-saturation (to the left of the curves; the two junctions are then forward) for low V_{CE}:

$V_{CB} = V_{CE} - V_{BE}$; in normal forward mode, if V_{CE} is very low, V_{BE} being positive, then V_{CB} can pass to negative: $V_C < V_B$; the base-collector junction passes to forward.

The two junctions are biased in direct: this is the saturation regime (of the base)!

1.4.3. *Early curves*

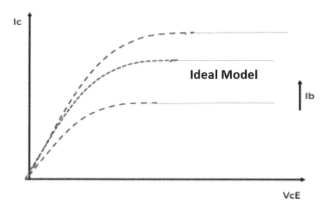

Figure 1.15. *(Ideal) Early curves*

NOTE.– The "epi." zone depends on collector doping and on the base-collector voltage. It limits the current term of the bipolar junction transistor and affects the maximum working frequency under strong currents.

– Base resistance r_b

The base resistance is very important for small signals (and noise, via its square). It varies a lot with the crowding effect – movement of minority carriers in the base toward its contacts, or tightening (focusing) of the equipotential ones at the level of this (these) contact(s) (and is difficult to measure because of r_E).

It essentially plays on the current gain and the collector current (see Early effect).

1.4.4. *Base width modulation; Early effect*

The width of neutral base W_B depends on the polarization of V_{BC}.

The Early effect, modulation of the neutral base width, affects the saturation current Is, the current gain b_F and the transit time t_{BF} through the base (therefore t_F, the total transit time).

This leads to a non-zero slope (V_A does not tend toward infinity). Its Early characteristics are represented by Ic (Vce).

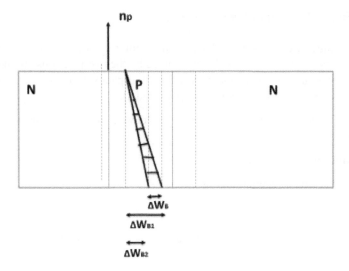

Figure 1.16. *Modulation of the base versus polarization*

Generally speaking, Early voltage can be defined as:

$$V_A^{-1} = \frac{1}{W_B(V_{BC}=0)} \cdot \frac{dW_B}{dV_{BC}} \qquad [1.48]$$

Very often, a linear variation of W_B is adopted as a function of V_{BC}.

$$V_A^{-1} = \frac{1}{W_B(0)} \cdot \frac{W_B(V_{BC})-W_B(V_{BC}=0)}{V_{BC}-0} \qquad [1.49]$$

$$W_B(V_{BC}) = W_B(0)\frac{W_B(V_{BC})-W_B(V_{BC}=0)}{V_{BC}-0}\left(1+\frac{V_{BC}}{V_A}\right) \qquad [1.50]$$

Thus, we have:

$$I_S(V_{BC}) = \frac{I_S(0)}{1+\frac{V_{BC}}{V_A}} \sim I_S(0)\left(1-\frac{V_{BC}}{V_A}\right) \qquad [1.51]$$

$$\beta_F(V_{BC}) = \frac{\beta_F(0)}{1+\frac{V_{BC}}{V_A}} \sim \beta_F(0)\left(1-\frac{V_{BC}}{V_A}\right) \qquad [1.52]$$

$$\tau_{BF} = \tau_{BF}(0)\left(1-\frac{V_{BC}}{V_A}\right)^2 \qquad [1.53]$$

1.4.5. *Ebers–Moll model wide signals*

The storage of charges in the base of the bipolar is modeled by the introduction of three types of capacitors. The loads induced by the two junctions are independent (same under saturation conditions). A diffusion capacity is associated with each of the two sources I_{cc} and I_{EC} (see the transport model):

– Capacitance (constant) insulating the device from the substrate (reverse junction).

– Diffusion capabilities.

– Neglected base recombinations.

– Low injection.

The Q_{DE} charges associated with I_{CC} are the sum of four minority charges.

Emitter side:

$$Q_{DE} = Q_E + Q_{JE} + Q_{BF} + Q_{JC} \qquad\qquad [1.54]$$

where:

– Q_{JE}: mobile minority charge in the base-emitter space load region associated with Icc (normally equal to 0);

– Q_E: mobile minority charge in the neutral emitter region;

– Q_{BF}: mobile minority charge stored in the neutral base region;

– Q_{JC}: mobile minority charge in the collector-base space charge region associated with Icc.

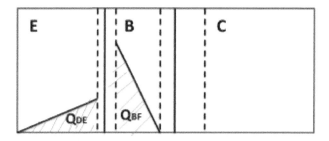

Figure 1.17. *B/E junction; distributed charges*

Since the total charge is zero, there will be as many minority charges as there are majority ones stored in each region. In order to determine the diffusion capacities, consideration will be given either to the minority or to the majority.

$$Q_{DE} = (\tau_E + \tau_{EB} + \tau_{BF} + \tau_{CB})\, I_{CC} = \tau_F \cdot I_{CC} \qquad [1.55]$$

where:

– t_{BF}: inverse of the cut-off frequency of the device.

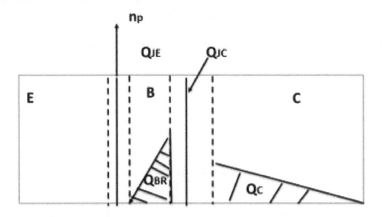

Figure 1.18. *B/C junction; diffused charges*

Q_{DC} charges associated with I_{EC} are the sum of four minority charges:

$$Q_{DC} = Q_C + Q_{JE} + Q_{BR} + Q_{JE} \qquad [1.56]$$

where:

– Q_{JE}: mobile minority charge in the collector-base charge space region associated with I_{Ec} (normally equal to 0);

– Q_C: mobile minority charge in the neutral collector region;

– Q_{BR}: mobile minority charge stored in the neutral base region;

– Q_{JE}: mobile minority charge in the collector-base charge space region associated with I_{EC} (normally equal to 0).

$$Q_{DC} = (\tau_C + \tau_{CB} + \tau_{BR} + \tau_{BE})\, I_{EC} = \tau_F I_{EC} \qquad [1.57]$$

where:

  – t_{CB}: ignored;

  – τ_C: delay in collector;

  – τ_{BR}: reverse transit time in the base;

  – τ_R: total reverse transit time.

In the saturated mode (V_{BE} and V_{BC} polarized directly), the charge stored in the bipolar junction is the sum of all these charges.

Q_{DE} and Q_{DC} constitute the total charge stored in the bipolar junction; they are independent. These two charges are modeled by nonlinear capacitances:

$$C_{DE} = \frac{dQ_{DE}}{dV_{BE})} = \frac{d(\tau_F I_{CC})}{dV_{BE}} \qquad [1.58]$$

$$C_{DC} = \frac{dQ_{DC}}{dV_{BC})} = \frac{d(\tau_R I_{EC})}{dV_{BC}} \qquad [1.59]$$

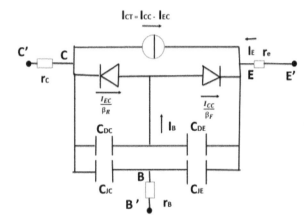

Figure 1.19. *Addition of capacitors in the "bipolar" model*

SPICE uses the following relationships:

$$C_{JE}(V_{BE}) = \frac{C_{JE}(0)}{(1 - V_{BE}/\emptyset_E)^{m_e}} \qquad [1.60]$$

And

$$C_{jC}(V_{BE}) = \frac{C_{jC}(0)}{(1 - V_{BC}/\emptyset_C)^{m_c}} \qquad [1.61]$$

m_e and m_c depend on the shape of junction profiles.

Φ_E and Φ_C are, respectively, the emitter-base and collector-base (diffusion) barrier potentials.

We obtain Q_{JE} and Q_{JC} charges by integrating on the areas of respective space charge with respect to applied voltages:

$$\left(Q_{JE}\right) = \int_0^{V_{BE}} C_{IE}.dV = \frac{C_{IE}(0)}{(1-m_e)}\left(1 - V_{BE}/\emptyset_E\right)^{1-m_e} \qquad [1.62]$$

$$\left(Q_{JC}\right) = \int_0^{V_{BC}} C_{IC}.dV = \frac{C_{IC}(0)}{(1-m_c)}\left(1 - V_{BC}/\emptyset_C\right)^{1-m_c} \qquad [1.63]$$

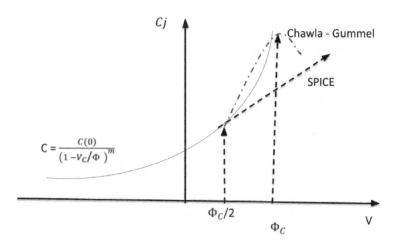

Figure 1.20. *Capacitors/junctions:*
Chawla–Gummel and "SPICE" model

For $V > F/2$, SPICE uses the following linear form:

$$C_J = 2^m C_{JE}(0)(2m\frac{V}{\emptyset} + (1-m)) \qquad [1.64]$$

This relationship is not very precise; but it should be noted, however, that it is the diffusion capacitances that predominate.

Nine new parameters are added to the Ebers–Moll model:

C_{je} (o), C_{jc} (o), F_E, F_C, F_S, t_F, t_R and F_C.

In integrated circuits, substrate capacitance is involved. It is a priori assumed to be constant, although it varies with the epitaxial substrate-layer potential. In fact, this is justified when this junction is reverse biased, which is the case when it serves as insulation.

1.4.6. *Current gain*

The reduction in the current gain b_F is due to additional components of I_B, which have been ignored up until now.

They are due to:

a) recomposition of carriers on surfaces;

b) recombination of carriers in the base-emitter space charge area;

c) formation of emitter-base surface currents.

As the variation of these three devices have the same shape as a function of V_{CE}, these are equivalent to a current of the following form:

$$I_B = I_{S,composite}(e^{\frac{qV_{BE}}{\eta_{EL}kT}} - 1) \qquad [1.65]$$

where n_{EL} is the low direct current emission factor; $1 < h_{EL} < 4$.

Most of the time, it is the device due to the recombination of carriers in the emitter-base space charge area that predominates (the other two can be minimized during the manufacturing processes).

$$I_B = I_{S,composite}\left(\exp\frac{qV_{BE}}{\eta_{EL}kt} - 1\right) + C_2 I_S(0)\left(\exp\frac{qV_{BE}}{\eta_{EL}kt} - 1\right)$$

$$+ \frac{Is(0)}{\beta_{BF}(0)}\left(\exp\frac{qV_{BC}}{\eta_{EL}kTkt} - 1\right) + C_4 I_S(0)\left(\exp\frac{qV_{BC}}{\eta_{EL}kt} - 1\right) \qquad [1.66]$$

$$C_4 = \frac{Is,c}{Is(0)}: \text{ in reverse.}$$

In this case, the drop in b is due to the effects of strong injection into the base (the base widens because the injection of carriers increases at the level of the base contact in order to "react" to the strong injection coming from the emitter; as the neutral base widens, t_B increases).

1.5. Simple bipolar junction transistor model

Common emitter assembly with resistive charge is given as:

Charge line equation: $V_{CC} = R_C \; I_C \; + \; V_{CE}$ [1.67]

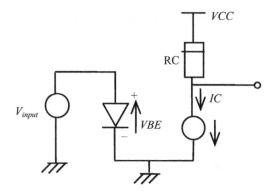

Figure 1.21. *Simple model of bipolar NPN in common emitter*

1.6. Network of static characteristics of the bipolar junction transistor

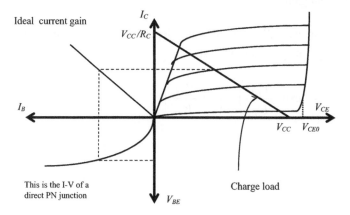

Figure 1.22. *Characteristic network of the NPN charge line*

$$I_C = \beta \, I_B \tag{1.68}$$

$$I_E = (\beta + 1) \, I_B$$

$$I_B = \frac{I_S}{\beta} \exp\left(\frac{V_{BE}}{k\,T\big/q}\right)$$

We write $U_T = \dfrac{k \cdot T}{q}$ with k = 1.38 × 10^{-23} J/K and q = 1.6 × 10^{-19} C

So, U_T ~ 26 mV at 300 K.

$$I_B = \frac{I_S}{\beta} \cdot \exp\left(\frac{V_{BE}}{U_T}\right) \tag{1.69}$$

$$I_C = I_S \cdot \exp\left(\frac{V_{BE}}{U_T}\right) \tag{1.70}$$

where:

– I_S: saturation current (proportional to n_i^2);

– β: current gain.

In reality, the gain β is not constant (dependence on the gain with V_{CE}) and the term V_A (Early voltage) is introduced:

$$I_C = I_S \cdot \exp\left(\frac{V_{BE}}{U_T}\right) \cdot \left(1 + \frac{V_{CE}}{V_A}\right) \tag{1.71}$$

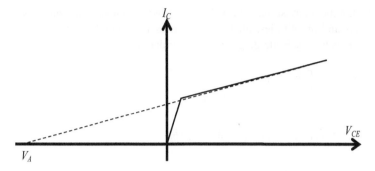

Figure 1.23. *Early voltage*

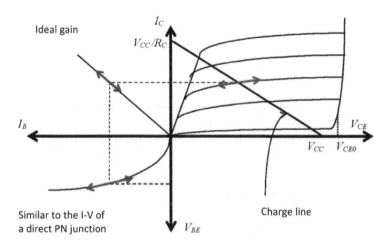

Figure 1.24. *Characteristic network (double arrows: vibrations superimposed on continuous characteristic lines (with Early effect on Ic(V$_{CE}$))*

The "small signals" model takes into account the electrical behavior of the device around a polarization point. The electrical model is linearized around a point (quiescent point), which facilitates calculating transfer functions. In fact, devices of a circuit are supplied via voltages and continuous currents; this then allows the circuit to "work", for example, by amplifying variable signals, which are small with respect to these continuous bias values.

In fact, almost any signal can be broken down into a sum of sinusoids of variable frequencies and amplitudes, all the more so as this signal "moves away" from the sinusoid. This is why Bode diagrams are calculated.

$$\begin{bmatrix} v_{be} \\ i_c \end{bmatrix} = \begin{bmatrix} h_{11} & h_{12} \\ h_{21} & h_{22} \end{bmatrix} \cdot \begin{bmatrix} i_b \\ v_{ce} \end{bmatrix}$$

[1.72]

$$v_{be} = h_{11} \cdot i_b + h_{12} \cdot v_{ce} \cong h_{11} \cdot i_b$$

[1.73]

$$i_c = h_{21} \cdot i_b + h_{22} \cdot v_{ce}$$

[1.74]

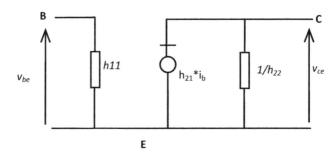

Figure 1.25. *NPN bipolar junction transistor with common emitter; hybrid parameters*

Transconductance is defined as:

$$g_m = \frac{1}{r_d} = \frac{\partial I_C}{\partial V_{BE}} = \frac{I_C}{U_T}$$

[1.75]

So $r_d = \dfrac{U_T}{I_C}$ (dynamic resistance)

NOTE.– r_d is a function of I_C.

$$\frac{1}{h_{11}} = \frac{\partial I_B}{\partial V_{BE}} = \frac{I_B}{U_T} = \frac{I_C}{\beta \cdot U_T} = \frac{g_m}{\beta} = \frac{1}{\beta \cdot r_d}$$

[1.76]

However, $r_\pi = h_{11} = \left[\dfrac{\partial I_B}{\partial V_{BE}} \right]^{-1}$

[1.77]

Therefore, $r_\pi = h_{11} = \dfrac{\beta}{g_m} = \beta \cdot r_d$ [1.77 bis]

$h_{21} = \beta$ [1.78]

Here, h_{22} considers the Early effect.

$$r_0 = \frac{1}{h_{22}} = \left[\frac{\partial I_C}{\partial V_{CE}}\right]^{-1} = \frac{V_A}{I_C}$$ [1.79]

Note that we have:

$$\beta \cdot i_b = \beta \frac{v_{be}}{h_{11}} = g_m \cdot v_{be}$$ [1.80]

Hence, the current usage is given as:

$$v_{be} = r_\pi \cdot i_b$$ [1.81]

$$i_c = \beta \cdot i_b + \frac{v_{ce}}{r_0} = g_m \cdot v_{be} + \frac{v_{ce}}{r_0}$$

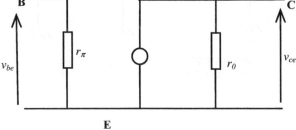

Figure 1.26. *Simplified equivalent diagram of the NPN bipolar junction transistor with a common transmitter*

1.6.1. *Common emitter configuration*

Collector charge: R_C

$$I_C = I_S \cdot \exp\left(\frac{V_{BE}}{U_T}\right)$$ [1.82]

$$I_C = I_S \cdot \exp\left(\frac{V_{input}}{U_T}\right) \qquad\qquad [1.83]$$

$$I_C = \beta \cdot I_B \qquad\qquad [1.84]$$

$$V_{output} = V_{cc} - R_C \cdot I_C \qquad\qquad [1.85]$$

$$V_{output} = V_{cc} - R_C \cdot I_S \cdot \exp\left(\frac{V_{input}}{U_T}\right) \qquad\qquad [1.86]$$

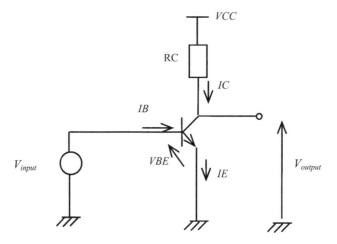

Figure 1.27. *Diagram of the NPN bipolar junction transistor with a common emitter*

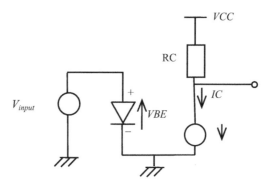

Figure 1.28. *NPN bipolar junction transistor with a common emitter*

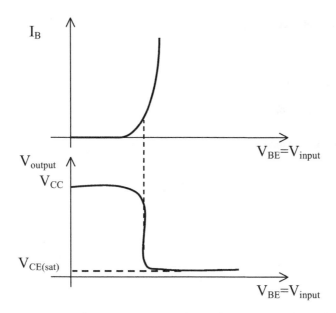

Figure 1.29. *Bipolar: an inverter*

The equivalent diagram can be obtained by short circuiting all the voltage sources.

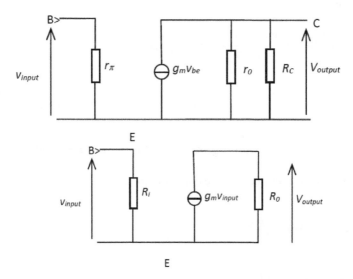

Figure 1.30. *Equivalent diagram of the bipolar junction transistor with a common emitter*

– Input resistance:

$$R_i = r_\pi \tag{1.87}$$

– Output resistance:

$$R_0 = \{ r_0 \,/\!/\, R_C \} \tag{1.88}$$

– Transconductance:

$$G_m = \frac{i_{output}}{v_{input}} = \frac{g_m \cdot v_{input}}{v_{input}} = g_m = \frac{\beta}{r_\pi} \tag{1.89}$$

– Voltage gain:

$$G_v = \frac{v_{output}}{v_{input}} = -\frac{g_m \cdot v_{input} \cdot R_0}{v_{input}} = -g_m \cdot R_0 = -\frac{\beta \cdot R_0}{r_\pi} \tag{1.90}$$

G_v is maximum for R_0 maximum or R_C "infinite". Then

$$G_{v\,max} = -g_m \cdot r_0 \tag{1.91}$$

Current gain:

$$G_i = \frac{i_{output}}{i_{input}} = \frac{g_m \cdot v_{input}}{v_{input} \Big/ r_\pi} = g_m \cdot r_\pi = \beta \tag{1.92}$$

This assembly makes it possible to have a voltage gain and current gain (> 1).

1.6.2. Common emitter configuration with emitter degeneration

A resistor R_E is introduced at the emitter output.

$$I_C = I_S \cdot \exp\left(\frac{V_{BE}}{U_T}\right) \tag{1.93}$$

$$V_{BE} = V_{input} - R_E \cdot I_E$$

R_E allows for getting a feedback on the basis. Indeed, if I_C increases, then the voltage drop at the terminals of R_E increases; consequently, V_{BE} decreases, leading to a decrease in I_B and therefore I_C (self-regulation).

1.7. Some applications

1.7.1. Current mirrors

We are trying to get $I_{C2} = I_{REF}$.

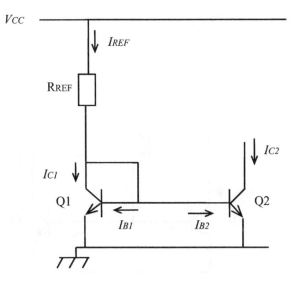

Figure 1.31. Current mirror

$$V_{BE1} = V_{BE2} \Rightarrow I_{B1} = I_{B2} \Rightarrow I_{C1} = I_{C2}$$

$$I_{REF} = I_{C1} + I_{B1} + I_{B2}$$

$$I_{REF} = I_{C1} + 2 \cdot \frac{I_{C1}}{\beta} = I_{C1} \cdot \left(1 + \frac{2}{\beta}\right)$$

$$I_{C1} = I_{C2} = \frac{I_{REF}}{1 + \frac{2}{\beta}} \cong I_{REF} = \frac{V_{CC} - V_{BE}}{R_{REF}} \qquad [1.94]$$

For $\beta = 100$, we get $I_{C1} = 0.98\ I_{REF}$.

– Improvement of the previous diagram

$$I_{E3} = I_{B1} + I_{B2} = 2\ I_{B2} = 2\ \frac{I_{C2}}{\beta} \qquad [1.95]$$

$$I_{B3} = \frac{I_{C3}}{\beta} = \frac{I_{E3}}{\beta + 1} = \frac{2\ I_{C2}}{\beta\ (\beta + 1)} \qquad [1.96]$$

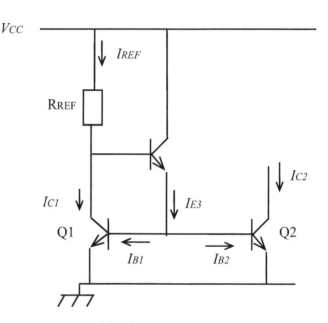

Figure 1.32. *Improved current source*

$$I_{REF} = I_{B3} + I_{C1} = I_{B3} + I_{C2} = \frac{2\ I_{C2}}{\beta\ (\beta + 1)} + I_{C2} \qquad [1.97]$$

Therefore,

$$I_{C2} = I_{C1} = \frac{I_{REF}}{1 + \dfrac{2}{\beta + \beta^2}} \qquad [1.98]$$

For $\beta = 100$, we get $I_{C1} = 0.9998\ I_{REF}$.

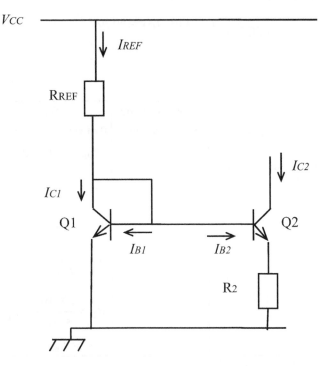

Figure 1.33. *Other configuration: Widlar source*

$$V_{BE1} - V_{BE2} = R_2 \cdot I_{E2} \cong R_2 \cdot I_{C2} \qquad\qquad [1.99]$$

$$I_{C1} = I_S \cdot \exp\left(\frac{V_{BE1}}{U_T}\right) \text{ and } I_{C2} = I_S \cdot \exp\left(\frac{V_{BE2}}{U_T}\right) \qquad\qquad [1.100]$$

therefore $V_{BE1} - V_{BE2} = U_T \cdot \ln\left(\dfrac{I_{C1}}{I_{C2}}\right)$ \qquad\qquad [1.101]

and finally $I_{C2} = \dfrac{U_T}{R_2} \cdot \ln\left(\dfrac{I_{C1}}{I_{C2}}\right)$ \qquad\qquad [1.102]

Comparison:

We are trying to get $I_{C1} = 5\ \mu A$. For this, we assume that the size of Q1 is 10 times larger than that of Q2 => $I_{C1} = 10 \cdot I_{C2} = 50\ \mu A$.

$V_{CC} = 5\ V$

"Simple" mirror	"Improved" mirror	"Widlar" mirror
$R_{REF} \approx V_{CC} \big/ I_{REF} = 10^5\ \Omega\ !!$ A resistor of such high value is difficult to integrate.	"Simple" mirror	By fixing: $R_{REF} = 10^4\ \Omega,$ $I_{C1} \cong \dfrac{V_{CC} - V_{BE}}{R_{REf}} = \dfrac{5 - 0,5}{10^4}$ $I_{C1} \cong 450\ \mu A$ $R_2 = \dfrac{U_T}{I_{C2}} \cdot \ln\left(\dfrac{I_{C1}}{I_{C2}}\right)$ $R_2 = \dfrac{26 \cdot 10^{-3}}{5 \cdot 10^{-6}} \cdot \ln\left(\dfrac{450}{5}\right)$ $R_2 \sim 23.4\ k\Omega$

1.7.2. Differential pair

$$I_{C1} = I_{S1} \cdot \exp\left(\frac{V_{BE1}}{U_T}\right) \qquad [1.103]$$

$$I_{C2} = I_{S2} \cdot \exp\left(\frac{V_{BE2}}{U_T}\right) \qquad [1.104]$$

$$V_{i1} - V_{BE1} + V_{BE2} - V_{i2} = 0$$

$$V_{i1} - V v_{i2} = V_{BE1} - V_{BE2} = V_{id}$$

$$I_{EE} = I_{E1} + I_{E2} = \frac{I_{C1} + I_{C2}}{\alpha}$$

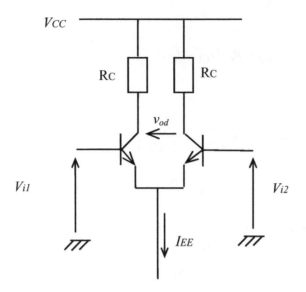

Figure 1.34. *Differential pair*

We can demonstrate that:

$$\frac{I_{C1}}{I_{C2}} = \exp\left(\frac{v_{id}}{U_T}\right)$$ [1.105]

$$I_{C1} = \frac{\alpha \cdot I_{EE}}{1 + \exp\left(\dfrac{v_{id}}{U_T}\right)}$$ [1.106]

$$I_{C2} = \frac{\alpha \cdot I_{EE}}{1 + \exp\left(-\dfrac{v_{id}}{U_T}\right)}$$ [1.107]

Additionally,

$$V_{O1} = V_{CC} - R_C \cdot I_{C1} \text{ and } V_{O2} = V_{CC} - R_C \cdot I_{C2}$$

that is,

$$V_{od} = V_{O1} - V_{O2} = R_C \cdot (I_{C1} - I_{C2})$$

Finally,

$$v_{od} = \alpha \cdot I_{EE} \cdot R_C \cdot \tanh\left(-\frac{v_{id}}{2 \cdot U_T}\right)$$ [1.108]

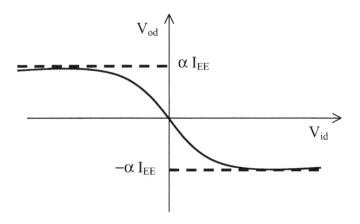

Figure 1.35. *Differential voltage gain*

The application of this assembly is differential amplification. The assembly can be assumed to be linear if $V_{id} \leq 10\ mV$, then

$$V_{od} = -\frac{\alpha \cdot I_{EE} \cdot R_C}{2 \cdot U_T} \cdot V_{id}$$ [1.109]

NOTE.– To widen the linearity range, emitters can be "degenerated", that is, by introducing an emitter resistance (feedback).

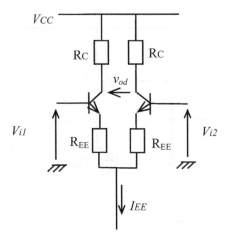

Figure 1.36. *Differential amplifier with emitter resistor*

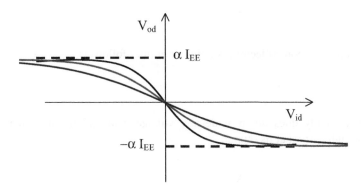

Figure 1.37. *An emitter resistance widens the linearity range*

1.7.3. Output stage

If $V_{input} > 0$, then Q1 is on (common collector assembly or emitter follower).

Q2 is off.

$$=> V_{output} \cong V_{CC}$$

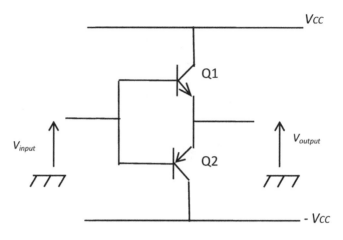

Figure 1.38. *Push-pull*

If $V_{input} < 0$, then Q1 is off.

Q2 is on (common collector assembly or emitter follower).

$$=> V_{output} \cong - V_{CC}$$

This assembly does not provide any voltage gain; on the other hand, it can deliver a high current at its output.

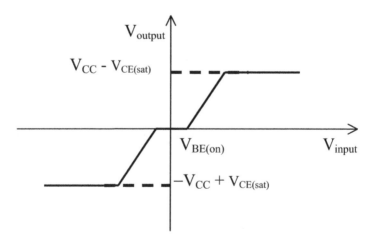

Figure 1.39. *Push-pull V_{input} versus V_{output}*

1.8. Application: operational amplifier

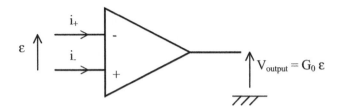

Figure 1.40. *Ideal operational amplifier (booster):* $i_+ = i_- = 0$ *and* $\varepsilon = 0$

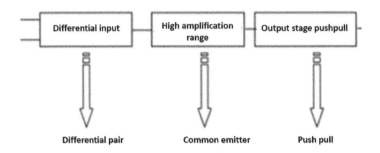

Figure 1.41. *Stages of the operational amplifier*

1.9. BiCMOS

BIPOLAR TECHNOLOGY (CMOS compatible)

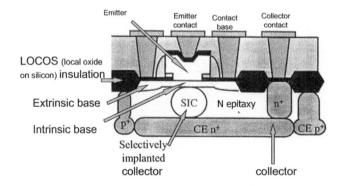

Figure 1.42. *BiCMOS technology*

MOSFET

2.1. Introduction

2.1.1. *Base structure*

The metal oxide semiconductor field effect transistor (MOSFET) consists of two highly doped zones/reservoirs, called source and drain (semiconductor regions of the same doping type), connected to their respective electrodes. Source and drain are produced in a region (sometimes directly the substrate) of the type opposite the substrate. A control electrode called a gate overhangs the channel area of the MOSFET.

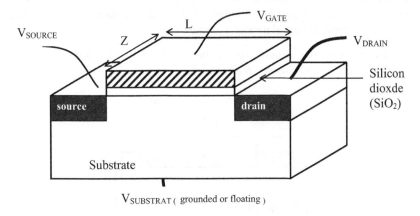

Figure 2.1. *Schematic structure of the MOS transistor*

For a color version of all figures in this chapter, see http://www.iste.co.uk/gontrand/analog. zip.

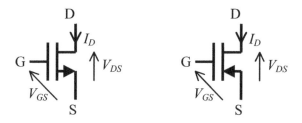

Figure 2.2. *NMOS symbol and PMOS symbol*

The MOS transistors have a unipolar current of majority carriers. The arrow in the gate source branch indicates the direction of the current (+ —> –):

– electrons in the case of the NMOS transistor (N-type source and drain);

– holes in the case of the PMOS transistor (P-type source and drain).

2.1.2. *Working principle*

1) The gate (G) makes it possible to control the current between the two source (S) and drain (D) electrodes (such as a water pipe whose flow rate would be controlled by applying more or less pressure).

2) Control is via the potential applied between the gate and source. This system, which consists of a MOS (metal-oxide-silicon) capacitor, enables control with very low current consumption. These are referred to as insulated gate transistors, where a displacement current can travel through the gate oxide.

3) The drain–source system can be likened to a resistor whose value varies between R_{on} (min value: closed circuit) and R_{off}.

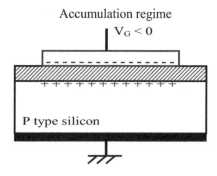

Figure 2.3. *Accumulation of majority carriers on the surface of silicon (enhancement regime)*

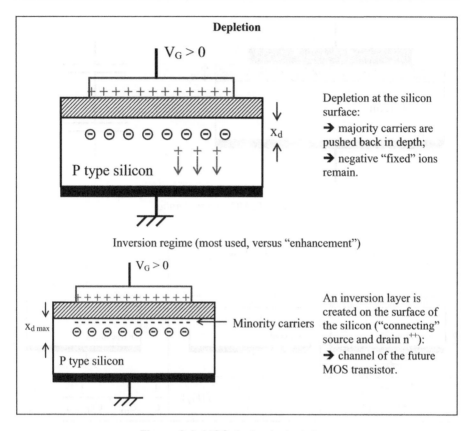

Figure 2.4. *MOS: technological steps*

2.2. MOS capability: electric model and curve C(V)

Electric model of the MOS capacitor (see the control grid – analogy with a valve and its seal: oxide).

MOS capacitance is equivalent to two capacitances in series: the oxide capacitance (C_{ox}) and the semiconductor capacitance (C_{SC}).

The equivalent total capacitance (per unit area) is written as:

$$\frac{1}{C} = \frac{1}{C_{ox}} + \frac{1}{C_{SC}}$$

[2.1]

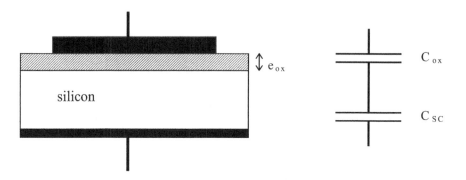

Figure 2.5. *MOS capacitance*

Accumulation	Depletion	Inversion
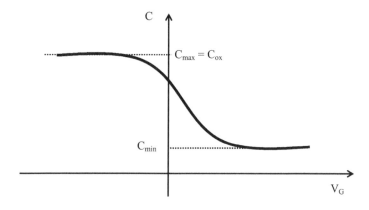		
$C_{SC} \gg C_{ox}$ $C \approx C_{ox} = C_{max}$	$\dfrac{1}{C} = \dfrac{1}{C_{ox}} + \dfrac{1}{C_{SC}} = f(V_G)$	$\dfrac{1}{C} = \dfrac{1}{C_{ox}} + \dfrac{1}{C_{SC,min}} = \dfrac{1}{C_{min}}$

Table 2.1. *Capacitance depending on operation mode*

Figure 2.6. *A C(V)*

2.3. Different types of MOS transistors

	N channel	P channel
Enhancement	Transistor normally off. Formation of a conductive N-channel by performing the inversion condition on the surface of the P-type material.	Transistor normally off. Formation of a conductive P-channel by performing the inversion condition on the surface of the N-type material.
Depletion	Transistor normally on. Suppression of conductive N channel by carrying out depletion.	Transistor normally on. Suppression of conductive N channel by carrying out depletion.

Table 2.2. NMOS and PMOS

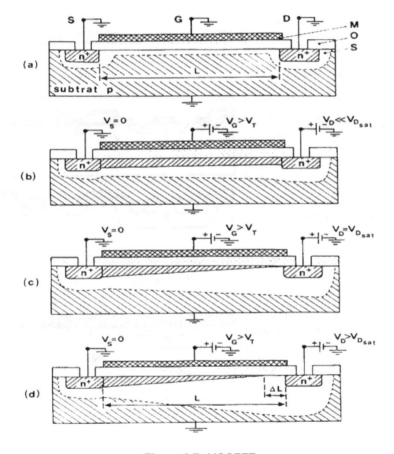

Figure 2.7. MOSFET

2.4. A CMOS technological process

We present a simple CMOS technological process (i.e. making it possible to jointly produce NMOS and PMOS transistors in the same substrate), with a single well.

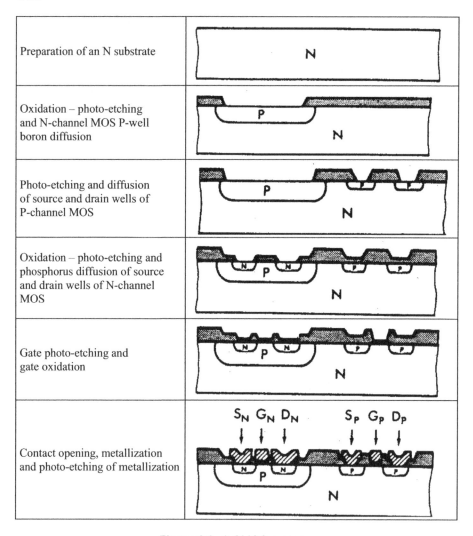

Figure 2.8. *A CMOS process*

Technology improvement (early 1970s): self-aligned gate, the source and drain areas (created by ion implantation) are created after the deposition of the polysilicon gate (serving as a mask).

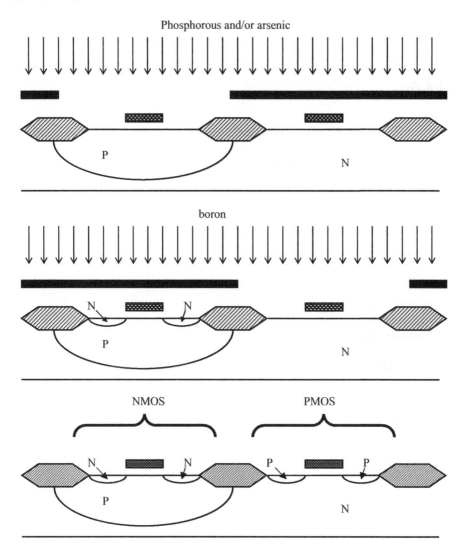

Figure 2.9. *Another CMOS process*

2.5. Electric modeling of the NMOS enhancement transistor

We deal hereafter with the different electric functional zones.

2.6. Off state

$V_G \ll V_{TH}$ (V_{TH}: threshold): there is no created channel, then $I_D = 0$, no matter what V_D.

2.7. Linear or ohmic or unsaturated regime

We assume the preformed channel:

$$I_D = \frac{V_{DS}}{R_{canal}} = \frac{V_{DS}}{\rho L} Z e \quad R = \rho \frac{L}{S} \tag{2.2}$$

$$I_D = \mu \frac{Z}{L} q n e V_{DS} \quad \rho = \frac{1}{n \mu q} \tag{2.3}$$

$Q = CV_G = e_{oxide}/e_{oxide.} ZLV_G$

$Surface = Z L$

$$\Rightarrow \frac{Q}{ZL} = C_{ox} V_G$$

$Q = q n Z L e$

$$\Rightarrow \frac{Q}{ZL} = q n e$$

$$I_D = \mu \frac{Z}{L} C_{ox} V_G V_{DS} \tag{2.4}$$

In reality, it is necessary to create the inversion channel, hence the notion of threshold voltage V_T (T: threshold), and the expression of the current is given as:

$$I_D = \mu \frac{Z}{L} C_{ox} \left(V_G - V_T \right) V_{DS} = \frac{V_{DS}}{R_{MOS}} \tag{2.5}$$

– Wide signal model:

with: $R_{MOS} = \dfrac{L}{\mu\, Z\, C_{ox}\left(V_G - V_T\right)}$ [2.6]

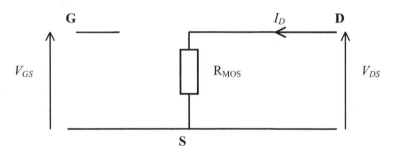

Figure 2.10. *Electric diagram of the MOS*

2.7.1. Saturation regime

For $V_{DS} = V_{DS(sat)}$, the channel is pinched; from this value of V_{DS}, the density of carriers arriving on the drain tends toward a constant.

For $V_G > V_T$ and V_{DS} close to $V_G - V_T$, we have:

$$I_D = \mu \frac{Z}{L} C_{ox} \left[\left(V_G - V_T\right)V_{DS} - \frac{V_{DS}^2}{2} \right]$$ [2.7]

2.7.2. High saturation velocity

For $V_G > V_T$ and high V_{DS}, we have:

$$I_D = \mu \frac{Z}{2 \cdot L} C_{ox} \left(V_{DS(sat)}\right)^2 \approx \mu \frac{Z}{2 \cdot L} C_{ox} \left(V_G - V_T\right)^2$$ [2.8a]

By taking an additional degree of refinement:

$$I_{D(sat)} \approx \mu \frac{Z}{2 \cdot L} C_{ox} \left(V_G - V_T\right)^2 \cdot \left(1 + \lambda \cdot V_{DS}\right)$$ [2.8b]

Draw (I_D) ½; it is a straight line!

– Wide signal model

$$I_{MOS} \approx \mu \frac{Z}{2 \cdot L} C_{ox} \left(V_G - V_T \right)^2 \qquad [2.9]$$

and $R_{MOS} = \dfrac{2L}{\lambda Z \mu C_{ox} \left(V_G - V_T \right)^2} \qquad [2.10]$

2.7.3. Static characteristics

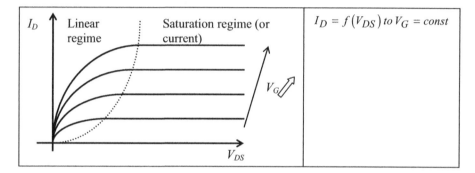

Table 2.3. $I_D(V_{DS})$

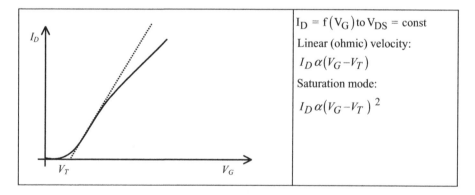

Figure 2.11. $I_D(V_{GS})$

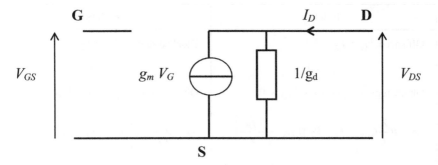

Figure 2.12. *Electric diagram of the MOS with its current source*

g_m: transconductance $g_m = \left. \dfrac{\partial I_D}{\partial V_G} \right|_{V_{DS}=const}$ [2.11]

g_d: channel conductance $g_d = \left. \dfrac{\partial I_D}{\partial V_D} \right|_{V_G=const}$ [2.12]

Ohmic (linear) velocity	Saturation mode
$g_m = \mu \dfrac{Z}{L} C_{ox} V_{DS}$ [2.13]	$g_m = \mu \dfrac{Z}{L} C_{ox} (V_G - V_T)$ [2.14]
	because $\lambda \cdot V_{DS} \ll 1$
	$g_m = \sqrt{\mu \dfrac{2Z}{L} C_{ox}} \ \sqrt{I_D}$ [2.15]

Table 2.4. *Transconductances versus electric regimes*

A comparison of the electric models of NMOS and PMOS transistors is presented in Table 2.5.

PMOS	NMOS
– Off speed: $V_G > V_T$	– Blocked regime: $V_G < V_T$
$I_D = 0$	$I_D = 0$
– Ohmic regime: $V_G - V_T < V_{DS} < 0$	– Ohmic regime: $V_G - V_T > V_{DS} > 0$
$I_D = -\mu\dfrac{Z}{L}C_{ox}\left[\,(V_G - V_T)V_{DS} - \dfrac{V_{DS}^2}{2}\,\right]$	$I_D = \mu\dfrac{Z}{L}C_{ox}\left[\,(V_G - V_T)V_{DS} - \dfrac{V_{DS}^2}{2}\,\right]$
– Saturated regime: $V_{DS} < V_G - V_T < 0$	– Saturated regime: $V_{DS} > V_G - V_T > 0$
$I_{D(sat)} \approx -\mu\dfrac{Z}{2 \cdot L}C_{ox}(V_G - V_T)^2 \cdot$ $(1 + \lambda \cdot V_{DS})$	$I_{D(sat)} \approx \mu\dfrac{Z}{2 \cdot L}C_{ox}(V_G - V_T)^2 \cdot$ $(1 + \lambda \cdot V_{DS})$

Table 2.5. *Drain current electric models of NMOS and PMOS transistors*

2.8. Applications

2.8.1. *Digital inverter*

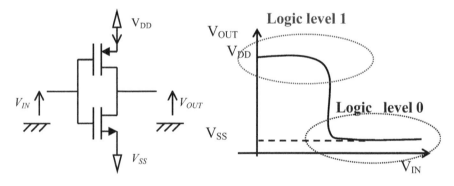

Figure 2.13. *CMOS inverter amplifier*

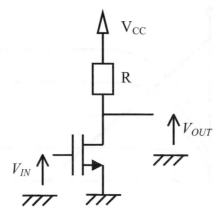

Figure 2.14. *A MOS amplifier*

Circuit equation: $V_{OUT} = V_{CC} - R I_D$

$- V_{IN} < V_T => I_D = 0 => V_{OUT} = V_{CC}$

$- V_{IN} > V_T$ at first $V_{DS} = V_{OUT} = V_{CC}$

$- V_G - V_T < V_{DS} =>$ MOSFET in saturation

$- I_D = \mu \dfrac{Z}{2 \cdot L} C_{ox} \left(V_G - V_T\right)^2$

$$V_{OUT} = V_{CC} - R I_D = V_{CC} - R \cdot \mu \dfrac{Z}{2 \cdot L} C_{ox} \left(V_G - V_T\right)^2$$

– Transition to a linear region

$V_{OUT} < V_G - V_T$

$I_D = \mu \dfrac{Z}{L} C_{ox} \left(V_G - V_T\right) V_{DS}$

$$V_{OUT} = V_{CC} - R I_D = V_{CC} - R \cdot \mu \dfrac{Z}{L} C_{ox} \left(V_G - V_T\right) V_{OUT}$$

$$V_{OUT(\min)} = V_{CC} - R I_{D(\max)} = V_{CC} - R \cdot \mu \dfrac{Z}{L} C_{ox} \left(V_{CC} - V_T\right) V_{OUT(\min)}$$

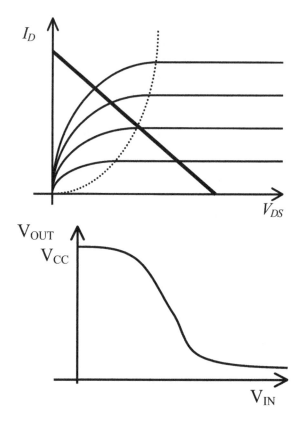

Figure 2.15. *Electric characteristics of the MOS: inverter*

2.8.2. *Active resistor*

A resistor is difficult to integrate; it is replaced by an active resistor produced by an NMOS or PMOS transistor whose gate is connected to the drain.

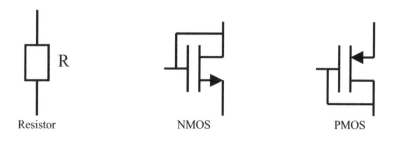

$V_{DS} = V_G > V_G - V_T$ → saturation mode

$$I_D = \pm \mu \frac{Z}{2 \cdot L} C_{ox} (V_G - V_T)^2 = \pm \mu \frac{Z}{2 \cdot L} C_{ox} (V_{DS} - V_T)^2 \qquad [2.16]$$

$$V_{DS} = V_T \pm \sqrt{\frac{2 |I_D| L}{\mu Z C_{ox}}} \qquad [2.17]$$

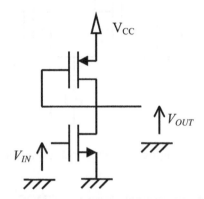

Figure 2.16. *Active charge AMP/MOS*

2.8.3. *MOS Single current mirror*

The calculations are developed in Volume 2 of this book.

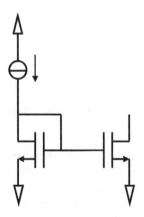

Figure 2.17. *Single current mirror*

2.8.4. *MOS differential amplifier*

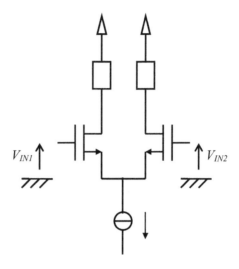

Figure 2.18. *MOS differential amplifier*

2.9. Explained technological steps of a CMOS

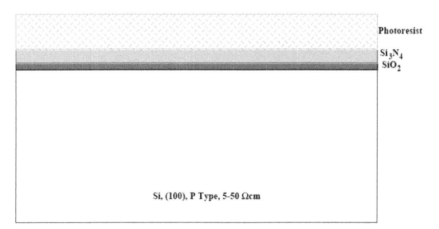

Substrate selection depends on the following points: moderately high resistivity, <100> orientation, P-type. Cleaning of wafers, thermal oxidation ($\approx$ 400 Å), Low Pressure Chemical Vapor Deposition (LPCVD) nitride deposit ($\approx$ 800 Å), photosensitive resin spin deposit ($\approx$ 0.5–1.0 µm) and curing.

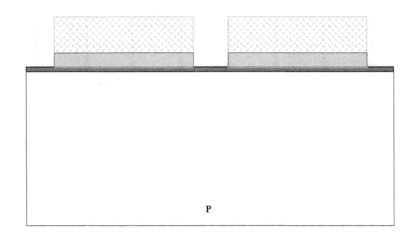

Mask 2.1. *Pattern of active areas.*
The nitride is dry-etched

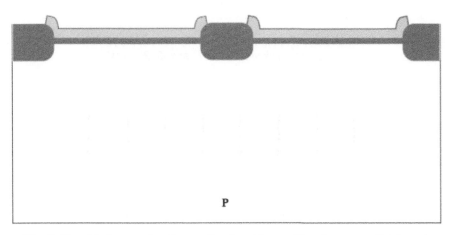

The field oxide is raw via a Locos (local oxide on silicon) process. It takes place at 90 min at 1,000°C in H_2O: growth $\approx$ 0.5 μm.

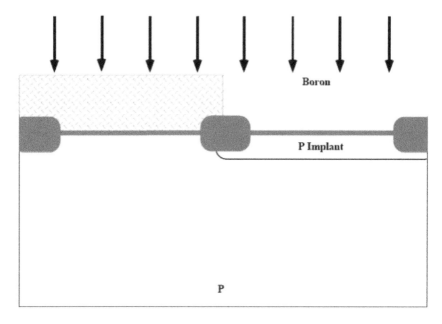

Mask 2.2. *An implant B+ forms the wells of the NMOS devices. Typically 10^{13} cm^{-2} at 150–200 keV*

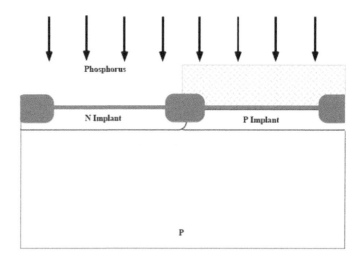

Mask 2.3. *A P+ implant to form the wells of the PMOS devices. Typically 10^{13} cm^{-2} at 300 keV*

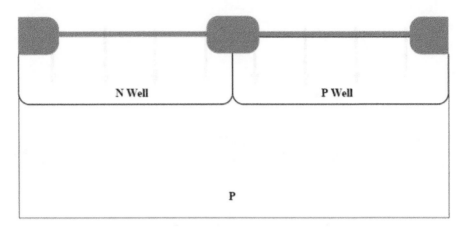

A drive-in (redistribution: dopants penetrate deeply) at high temperature controls the "final" well depths and repairs (in a few milliseconds) damage caused by the implantation. It usually takes 4–6 h at 1,000°C–1,100°C .

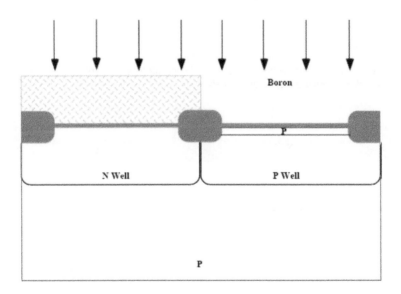

Mask 2.4. *Used to hide future PMOS. A V_{TH} adjustment implant (TH: thresold, threshold) is performed on NMOS devices, generally a 1.5 × 1.012 cm^{-2} boron implant at 50–75 keV*

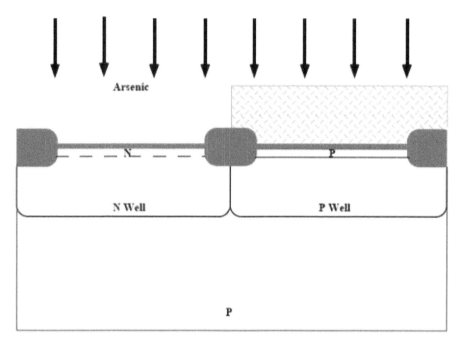

Mask 2.5. *Used to hide NMOS. A V_T adjustment implant is performed at PMOS devices; typically a 1–5 × 10^{12} cm^{-2} As implant at 75–100 keV*

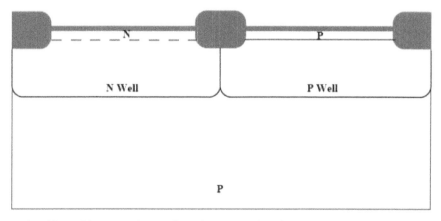

The thin oxide on active regions is removed and a new gate oxide is treated: from 50 to 100 Å, in 1–2 h at 800°C under O_2.

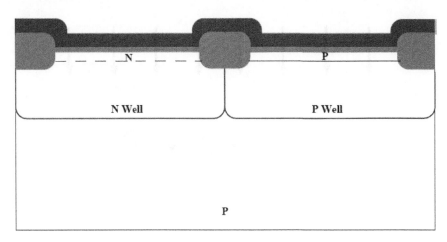

Poly. is deposited by LPCVD ($\approx$0.5 μm). A P implant dopes the poly gate (usually 5×101^5 cm^{-2}).

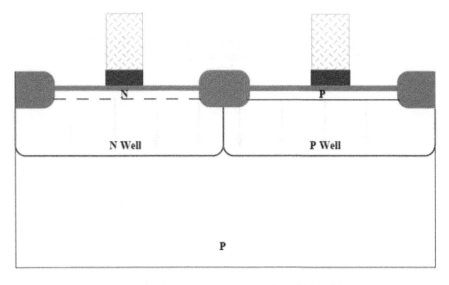

Mask 2.6. *Used to protect gates from MOS. The poly is plasma etched using anisotropic etching*

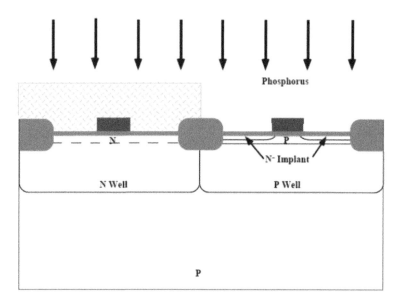

Mask 2.7. *Protects PMOS. A P implant forms the LDD (lightly doped drain) regions in the NMOS (generally 5 × 10^{13} cm^{-2} at 50 keV)*

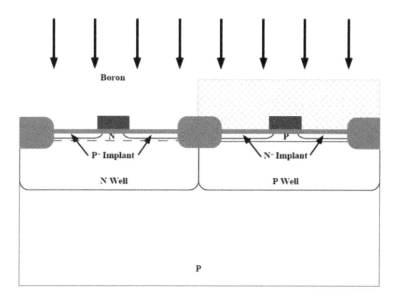

Mask 2.8. *Protects NMOS. An implant B forms the LDD regions in the PMOS (generally 5 × 10^{13} cm^{-2} at 50 keV)*

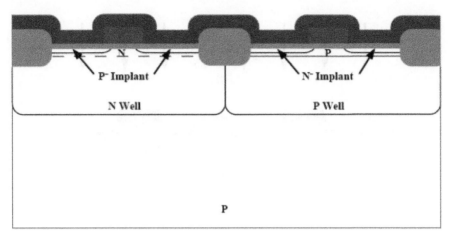

A conformal SiO$_2$ layer is deposited (typically 0.5 μm).

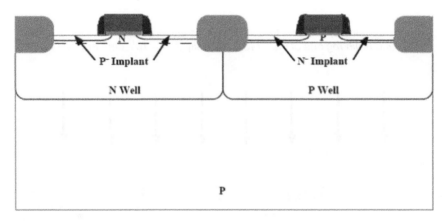

An anisotropic etching leaves "sidewalls", flanking the edges of the polysilicone gates.

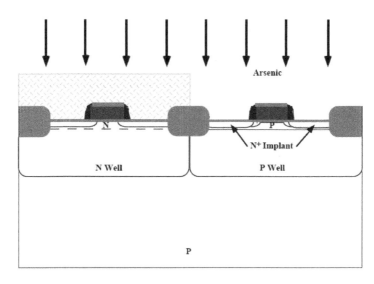

Mask 2.9. *Protects the PMOS; an As implant forms the source and drain regions of the NMOS (typically 2–4 × 10^{15} cm^{-2} at 75 keV)*

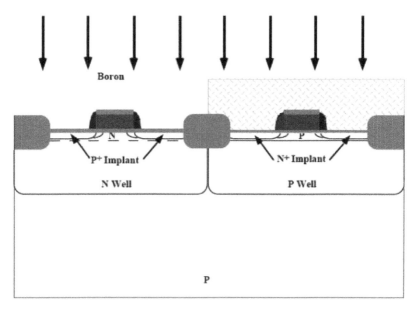

Mask 2.10. *Protects NMOS devices, implant B forms the source and drain of the PMOS (usually 1–3 × 10^{15} cm^{-2} at 50 keV)*

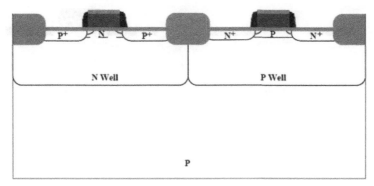

A final high-temperature annealing determines the depth of junctions and repairs implant damage (typically 30 min at 900°C or 1 min rapid thermal annealing (RTA) at 1,000°C).

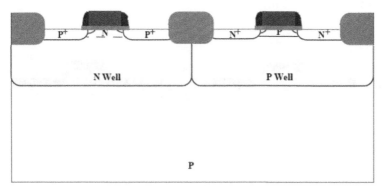

Oxide etching makes it possible to define the contacts with Si and poly-Si regions.

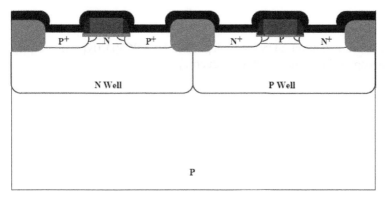

Some Ti (titanium) is deposited by sputtering (typically 1,000 Å).

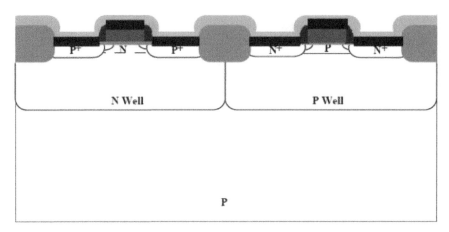

Ti reacts in an atmospheric (ambient) N_2 (nitrogen), forming $TiSi_2$ and TiN (typically 1 min a 600°C).

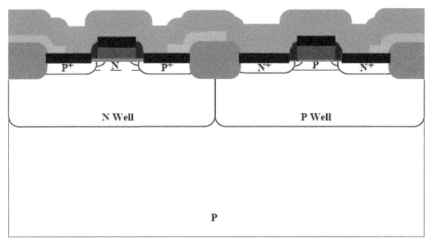

A conformal layer of SiO_2 is deposited by LPCVD (typically 1 µm, often a borophosphosilicate glass (BPSG) – ductile).

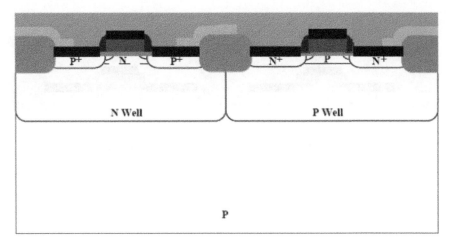

A CMP (chemical mechanical processing) is used to planarize the wafer's surface.

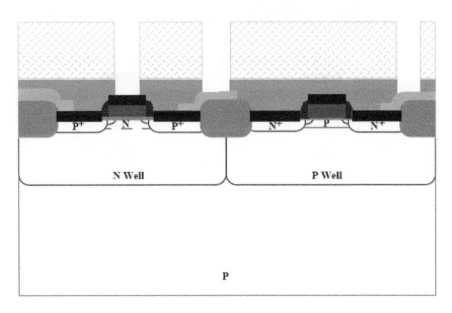

Mask 2.11. *Used to define contact holes. The SiO$_2$ is etched*

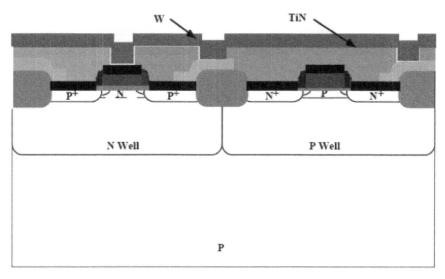

A thin barrier layer of TiN is deposited by sputtering (generally a few hundred Å), followed by CVD deposit of W (tungsten).

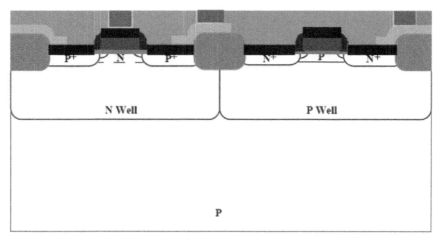

CMP is used to planarize the wafer's surface, completing the damascene process.

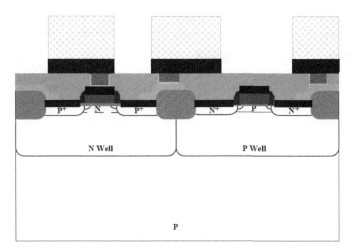

Al is deposited on the wafer by sputtering.

Mask 2.12. *Used to etch final Al pilar contacts*

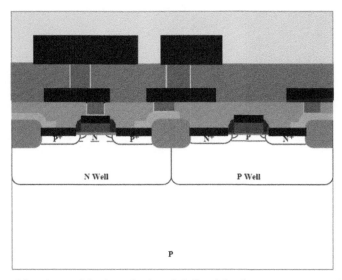

A second level metal is deposited and defined in the same way as Al level 1. Mask 14 is used to define the bias contact, and Mask 15 is used to define metal 2. A final passivation layer of Si_3N_4 (nitride) is deposited by PECVD (plasma enhanced (PE)) and defined by Mask 16.

Figure 2.19. *Explained technical steps of a CMOS*

2D MESHES (MOS)

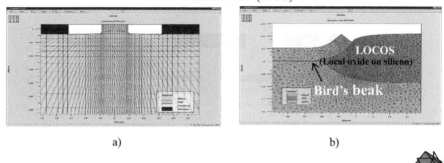

a)   b)

Figure 2.20. *2D meshes (MOS): (a) finite differences; (b) finite elements*

Devices Dedicated to Radio Frequency: Toward Nanoelectronics

3.1. Introduction

Silicon-Germanium (SiGe) heterojunction bipolar junction transistors (HBTs) appeared in order to satisfy the very high demand of telecommunications in terms of data speed and mass. Devices containing SiGe alloys are much faster than their fully silicon counterparts. In fact, the SiGe HBT, having an epitaxial base of SiGe, or even SiGeC, became very attractive for wireless and optical (digital) telecommunications; in this case, it is integrated in advanced BiCMOS (CMOS compatible bipolar) technology.

Faced with the limitations of bipolar junction transistors to reach high-frequency applications (typically 100 kHz–100 GHz), inserting germanium into the base was the easiest solution to implement in terms of performance/cost. The heterojunction provided by the SiGe material has made it possible to greatly improve the performances of silicon-based bipolar junction transistors and make them competitive with III/V compounds (AsGa type) for high-frequency applications. The appeal of the $Si_{1-x}Ge_x$ (IV–IV compound), compared to III–V materials, lies in the fact that the microelectronics industry is based on silicon (sand!); its technology is already very advanced, and its production cost is relatively low. Thus, the know-how of silicon technology is combined with the physical characteristics that the SiGe(C) alloy provides to increase the performance of bipolar junction transistors.

The fundamental physical advantage of $Si_{1-x}Ge_x$ material is that it has a smaller bandgap than Si. This property is exploited in the base of bipolar junction transistors

For a color version of all figures in this chapter, see http://www.iste.co.uk/gontrand/analog.zip.

to differentiate the forces and potential barriers applied to electrons and holes. This makes it possible to improve the efficiency of injecting electrons from the emitter, for an NPN, toward the base, while blocking holes as effectively as possible (reduction of Ib). The injection of electrons from the emitter to the base is therefore favored, which makes it possible to obtain a higher collector current. This results in a greater current gain for the SiGe HBTs.

3.2. Model for HBT SiGeC and device structure

3.2.1. *Modeling the drift–diffusion equation*

We can recall the fundamental equations for analyzing the electrical behavior of semiconductors; we base our analysis on a macroscopic description of semiconductors with a possible nonuniform composition. Various semiconductors differ in their fundamental properties such as the bandgap, the mobility of carriers, the effective masses of electrons and holes. In addition, interfaces between different materials must be properly described.

We re-write standard diffusion and drift equations to model the SiGeC alloy heterostructure. The Poisson equation considers that the dielectric constant is a function of position. In continuity equations, the Shockley–Read–Hall model is considered, as well as the position-dependent intrinsic carrier concentration.

– Poisson equation:

$$\nabla^2 \varphi = \frac{-q}{\varepsilon_{S/C}} \, [p - n + N_D^+ - N_A^-] \qquad [3.1]$$

– Continuity equations for electrons and holes:

$$\frac{\partial n}{\partial t} = GR_n + \frac{1}{q} \frac{dJ_n}{dx} \qquad [3.2]$$

$$\frac{\partial P}{\partial t} = GR_p - \frac{1}{q} \frac{dJ_p}{dx} \qquad [3.3]$$

Equations of current densities for electrons and holes are given as:

$$J_n = -qn\mu_n \frac{d\varphi_n}{dx} \qquad [3.4]$$

$$J_p = -q\, p\mu_p \frac{d\varphi_p}{dx} \tag{3.5}$$

– N_D and N_A are concentrations of ionized impurities; ε is the dielectric constant of the material and q is the charge bound to an electron or hole. Electron and hole current densities are functions of concentrations, mobilities μ_n and μ_p of carriers, and the Fermi quasi-potentials for the electron and the hole, Φn and Φp, respectively. Equations and physical models involved in simulating $Si_{1-x}Ge_x\ C_y$ x HBTs are applied in our finite difference 2D simulator.

First, the simulator solves the partial differential equations (PDEs) for the electrostatic potential Φ and for the concentration of electrons and holes, n and p, respectively. GRn and GRp are the net generation/recombination rates for holes and electrons, respectively. The recombination and generation models for SiGeC heterojunction are the same models previously described for homojunction; therefore, we use the well-known Shockley–Read–Hall (SRH) model.

$$GR_n = GR_p = \frac{n \cdot p - n_i^2}{\tau_n\left(p + p_i\right) + \tau_p\left(n + n_i\right)} \tag{3.6}$$

τ_n and τ_p are the lifetimes of electrons and holes in the semiconductor.

$$n_{i\ SiGeC}^2 = \left(N_C \cdot N_V\right)_{SiGec} \cdot \exp\left(-\frac{E_{g\ SiGec}}{KT}\right) \tag{3.7}$$

$$\varphi_n = -\frac{1}{q}E_{FN}\ ,\ \varphi_p = -\frac{1}{q}E_{FP} \tag{3.8}$$

$$E_{FN} = E_c + KT \ln\left(\frac{n}{N_c}\right) + KT \ln \gamma_n \tag{3.9}$$

$$E_{FP} = E_V - KT \ln\left(\frac{p}{N_v}\right) + KT \ln \gamma_p \tag{3.10}$$

$$\gamma_n = \gamma_p = 1 \text{ for Boltzmann statistics.} \tag{3.11}$$

$\gamma_n\ \gamma_p$ are parameters related to Fermi–Dirac statistics (low temperatures):

$$E_C = -q\,\varphi + \frac{E_g}{2} \qquad\qquad [3.12]$$

$$E_V = -q\,\varphi - \frac{E_g}{2} + \Delta E_v \qquad\qquad [3.13]$$

ΔE_V is the valence band discontinuity; it is not clear how this bandgap is distributed in the SiGeC material, but it appears that the gap (bandgap) propagates mainly in the valence band for SiGe.

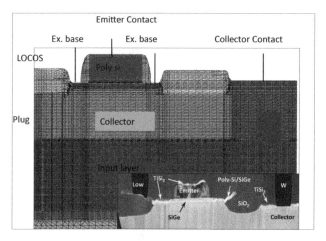

Figure 3.1. *Schematic cross-section of a heterojunction bipolar junction transistor, integrated in 0.13 μm BiCMOS technology with an aligned emitter (polysilicon) architecture*

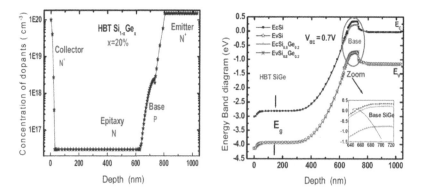

Figure 3.2. *Cross-section of a typical heterojunction bipolar transistor (HBT)*

3.2.1.1. *Electric results*

We studied the effect of little carbon 0.5%, 0.75%, and 1% in the SiGe HBT device. The addition of carbon to the base causes an increase in the base current. This increase is due to the reduction in the lifetime of the minority carriers in this base due to the low incorporation of carbon in the SiGeC, combined with the presence of deep traps in the emitter-base junction. The decrease in Ic can be explained by the widening of the bandgap and especially the increase in the boron dose in the neutral base region due to the reduction in the diffusion of the latter. An increase in I_B and a reduction in Ic induces a reduction in the current gain.

Figure 3.3 illustrates the evolution of the current gain versus Vbe. The more carbon there is in the device, the greater the decrease in current gain. The results obtained in this study are effectively compared with electric characteristics obtained by SPICE-type simulations (see Kirchhoff equations) via parameter extractions from simulations using compact models (ratio 5) implemented in commercial ADS (advanced design system) simulator standards. In Figures 3.4 and 3.5, Gummel characteristics and the current gain for a SiGeC HBT are shown, which are simulated by our digital simulator and the commercial ADS simulator, compared with HBT measurements with variable sections ($0.17 \times 1.9 \ \mu m^2$) and ($0.17 \times 19.9 \ \mu m^2$). We observe a good agreement between simulations and measurements. In addition, it should be noted that the maximum current is constant for emitter surfaces $0.3 \ \mu m^2$ to $1.600 \ \mu m^2$ (ratio 5.000). The maximum current gain is about 1.600 at Vbe = 0.6 V.

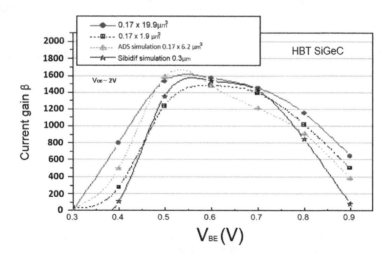

Figure 3.3. *Current gain versus V_{BE} for SiGeC HBT simulated by our digital software and compared to a commercial ADS simulator and measurement transistors*

Figure 3.5 illustrates Early characteristics, I_C-V_{CE}, for a "SiGeC" with a carbon concentration of 0.75%: HBT simulated by our digital simulator and the simulator and the commercial ADS (0.17×6.2 μm^2), compared for measurements on surfaces of different emitters (0.17×5.9 μm^2); again, we find a good agreement between the simulation and the measurements. The electric characteristics of this device may be penalized by the presence of defects inherent in the complex shrinkage of the structure.

For our devices, most of these suitable defects are found at the vertical interface between spacers and the polysilicon emitter, due to the reactive-ion etching process (RIE) step. Nevertheless, their location and their effective density or their capture section have an influence on the electrical characteristics of HBT. Figure 3.4 represents the Gummel characteristics for two similar SiGeC HBTs: one with hole traps placed at 0.6 eV (recall: mid-gap = 0.56 eV) in the semiconductor space, and with an effective density $N_T = 10^{18} cm^{-3}$ and a capture section σ $10^{-17} cm^2$ – according to our DLTS (deep level transient spectroscopy) measurements – and the other: a SiGeC HBT considered without defects.

These defects also induce noise phenomena, which are all the more prohibitive when we engage in nanoelectronics.

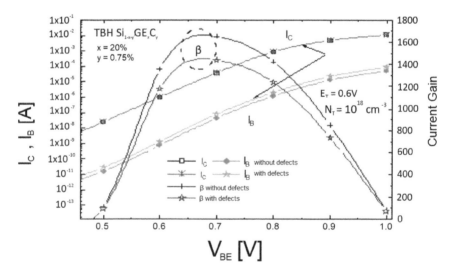

Figure 3.4. *Current gain versus V_{BE} for SiGeC HBT by our simulator compared to commercial ADS simulator and measurement transistors to various emitter sections*

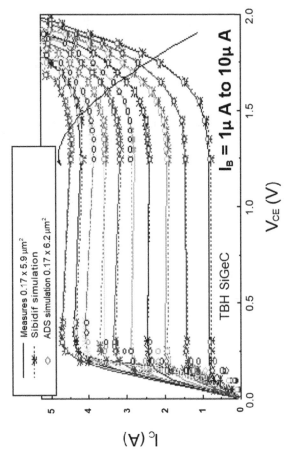

Figure 3.5. *Early curves; I_C–V_{CE} for simulated SiGeC HBT compared to the commercial ADS simulator (0.17 × 6.2 μm²) and the measured transistors (0.17 × 5.9 μm²)*

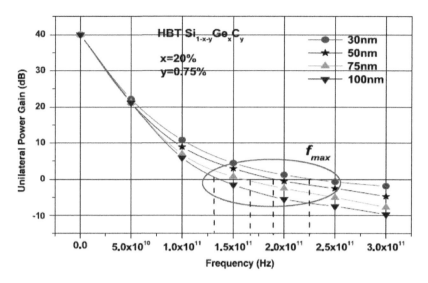

Figure 3.6. *Power gain as a function of frequency*

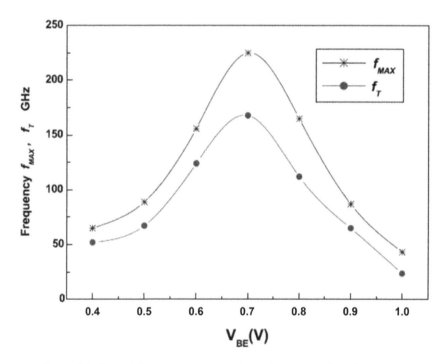

Figure 3.7. *Transit frequency and maximum frequency, function of Vbe*

3.3. MOS of the future?

3.3.1. *Introduction*

The future of microelectronics, in general and in particular that of the MOS transistors, depends on the ability of the industry and researchers to continue the race for integration through miniaturization. To continue this miniaturization, it is interesting to investigate new transistor architectures.

The introduction of new gates compared with conventional MOS transistors is beneficial for the operation of devices in several respects. The advantage of the SOI (silicon on insulator) structure is added to other improvements: double-gate transistors. The advantage of a double MOSFET gate is to control the silicon channel very efficiently with a very small channel width. It consists of applying a gate contact to both sides of the channel. This concept eliminates short channel effects and leads to a higher current compared to a conventional TMOS.

This has some advantages, in particular the improvement of the threshold characteristics of the transistors (increase of the I_{ON}/I_{OFF} ratio, increase of drain current or better still the mobility of carriers in the channel, etc.).

However, the double-gate transistor, like the single-gate transistor, has limitations with nanoscale dimensions. This miniaturization has a direct effect on the thickness of the gate oxide. Silicon oxide (SiO_2) reaches limit dimensions, hence the search for new architectures. The semiconductor industry, guided by the ITRS (International Technology Roadmap for Semiconductors), must be oriented toward new technologies.

"High-k" technology: this technology is based on the integration of new materials with high permittivity, making it possible to reduce the leakage current through the gate.

– Fully depleted silicon on insulator (FDSOI) technology: this technology is designed by CEA LETI and based on the integration of a thin layer of insulating silicon oxide with the conventional architecture of transistors. It is a planar process technology that offers the benefits of silicon's reduced geometries, while simplifying the manufacturing process.

This innovation gives the transistors efficient and energy-efficient operation, while continuing the challenge of miniaturization.

This technique is complex and therefore expensive. Nevertheless, companies such as STMicroelectronics seem to be fully committed to it.

In recent years, while the transistor has a size of less than a few tens of nanometers (node currently targeted at ST Crolles: 10 nm), the effort has been focused on the challenges of each new technological generation. An example is leakage current, which, because transistors are so small, now accounts for a significant proportion of its energy consumption.

In order to continue to deliver superior performance, while keeping leaks under control, silicon transistors have become increasingly complex, adding additional manufacturing steps and more recently considering moving to a new, expensive 3D architecture. The SOI approach (from SOITEC) therefore consists of interposing a thin insulating oxide layer (tens of angstroms), typically between an epitaxial layer and the silicon substrate; this replaces solation by reverse biased junction, with the presence of its Is.

3.3.2. *DGMOS*

In this section, we present an efficient digital model of DGMOS (double-gate MOS) based on a description via quantum mechanics. It provides a self-consistent solution to the Schrödinger and Poisson equations. Quasi-ballistic transport can also be considered.

Carrier potential and concentration were calculated using a program based on a scheme of finite differences with a uniform mesh. The Newton–Raphson method makes it possible to properly consider the coupling of the aforementioned equations.

3.3.2.1. *Structure*

Figure 3.8 illustrates a DGMOSFET (double gate MOSFET).

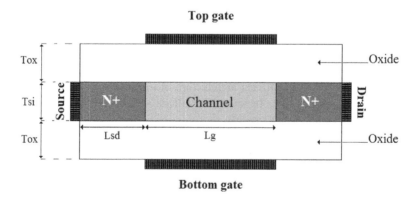

Figure 3.8. *Schematic diagram of a DGMOS transistor*

The transport of electrons takes place in the channel, which is the active region outside the two source reservoirs and the drain. We considered a DG MOSFET with the following parameters: doping level NA = 10^{10} cm^{-3}; source/drain doping N + = 10^{20}cm^{-3}; TSi thickness = 1.5 nm, oxide thickness TOX = 1.5 nm; channel length LG = 10 nm; source/drain length L$_{SD}$ = 5 nm. Metallization is carried out with an aluminum material; the working function is then considered to be 4.25 eV. Voltages V$_{GS}$ applied to the two gates are identical. All calculations were performed at room temperature (T = 300 K).

3.3.3. *Transport in nanoscale MOSFETs*

The transport of a "random" type of carriers (described in a conventional manner) considers collisions (with defects, phonons, etc.). Furthermore, the MOS transistor has reached nanometric dimensions and the transport of carriers in the active region (channel) is modified because the length of the channel is of the order of the mean distance between two interactions, called the mean free path. A large part of the carriers is then able to pass through the channel without suffering a collision. Electrons can move without diffusion as in a vacuum tube: in this case, the transport is called ballistic (like a projectile motion of a bullet).

3.3.3.1. *Electron density*

The energy profile in the ballistic channel MOS transistor is illustrated in Figure 3.9.

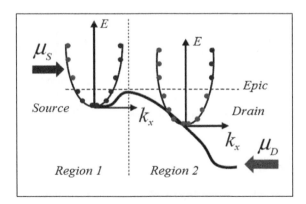

Figure 3.9. *Conduction band evolution, along the drain-source axis, for ballistic transport*

This profile can be divided into two regions: points to the left of the peak energy, Epic, (region 1) and points to the right of the peak energy (region 2). Region 1:

electrons of energy lower than Epic come from the source reservoir and electrons of energy higher than that of the Epic source are moved to the left, from the drain reservoir. The same explanation can be made in region 2. This is explained by the fact that each wave is associated with an incident wave and a reflected wave (quantum mechanics). The two reservoirs (source, drain) are characterized by two Fermi levels $\mu_S \wedge \mu_D$.

In region 1, electron density can be written as follows:

$$
\begin{aligned}
:n_{left}(x, E_{kj}) = &\int_0^\infty \left[\frac{1}{\pi\hbar}\sqrt{\frac{m_x^*}{2E_x}}\frac{1}{1+exp((E_x+E_i+E_{kj}-\mu_s)/k_BT)}\right]dE_x + \\
&\int_0^{E_{pic}}\left[\frac{1}{\pi\hbar}\sqrt{\frac{m_x^*}{2E_x}}\frac{1}{1+exp\left(\frac{E_x+E_i+E_{kj}-\mu_s}{k_BT}\right)}\right]dE_x + \\
&\int_{E_{pic}}^\infty\left[\frac{1}{\pi\hbar}\sqrt{\frac{m_x^*}{2E_x}}\frac{1}{1+exp((E_x+E_i+E_{kj}-\mu_D)/k_BT)}\right]dE_x
\end{aligned}
\tag{3.14}
$$

The left index is used for region 1.

Taking into account the contributions of all the transversal modes on the E_{kj} integration, we obtain:

$$
n_{left}(x) = \int_0^\infty \frac{1}{\pi\hbar}\sqrt{\frac{m_y^*}{2E_{kj}}}\left[n_{left}(x, E_{kj})\right]dE_{kj}
\tag{3.15}
$$

$$
\begin{aligned}
n_{left}(x) = \\
n_{2Di}\left\{\ln(1 + e^{\tilde{\mu}_s}) + \frac{1}{\sqrt{\pi}}\int_0^{\tilde{E}_{pic}}\frac{d\tilde{E}_x}{\sqrt{\tilde{E}_x}}\Im_{-\frac{1}{2}}(\tilde{\mu}_s - \tilde{E}_x) + \frac{1}{\sqrt{\pi}}\int_{\tilde{E}_{pic}}^\infty\frac{d\tilde{E}_x}{\sqrt{\tilde{E}_x}}\Im_{-\frac{1}{2}}(\tilde{\mu}_D - \tilde{E}_x)\right\}
\end{aligned}
\tag{3.16}
$$

where the cap ~ means that all the quantities are expressed with respect to the potential local sub-band $E_i(x)$ and normalized with respect to thermal energy k_BT and n_{2Di}; the electron density for each sub-band is given by:

$$
n_{2Di} = \frac{\sqrt{m_x^*m_y^*}}{\pi\hbar^2}\frac{k_BT}{2}
\tag{3.17}
$$

In the same way, electron density in region 2 can be obtained by:

$$
\begin{aligned}
n_{right}(x) = n_{2Di}\left\{\ln(1 + e^{\tilde{\mu}_D}) + \frac{1}{\sqrt{\pi}}\int_0^{\tilde{E}_{pic}}\frac{d\tilde{E}_x}{\sqrt{\tilde{E}_x}}\Im_{-1/2}(\tilde{\mu}_D - \tilde{E}_x) + \right. \\
\left.\frac{1}{\sqrt{\pi}}\int_{\tilde{E}_{pic}}^\infty\frac{d\tilde{E}_x}{\sqrt{\tilde{E}_x}}\Im_{-1/2}(\tilde{\mu}_s - \tilde{E}_x)\right\}
\end{aligned}
\tag{3.18}
$$

3.3.3.2. *Electron density*

Current is maintained throughout the device; it can be evaluated in a cross-section perpendicular to the direction of its flow. To keep a general result, we assume that the section is in x, which is to the left of the peak of the barrier. For each sub-band, the current density in ballistic regime is written as:

$$J(E_{kj}) =$$

$$\int_0^{E_{pic}} \left[\sqrt{\frac{2E_x}{m_x^*}} \frac{q}{\pi\hbar} \sqrt{\frac{m_x^*}{2E_x}} \frac{1}{1+\exp\left(\frac{E_x+E_i+E_{kj}-\mu_S}{k_BT}\right)} \right] dE_x \; +$$

$$\int_{E_{pic}}^{\infty} \left[\sqrt{\frac{2E_x}{m_x^*}} \frac{q}{\pi\hbar} \sqrt{\frac{m_x^*}{2E_x}} \frac{1}{1+\exp\left(\frac{E_x+E_i+E_{kj}-\mu_S}{k_BT}\right)} \right] dE_x \; -$$

$$\int_0^{E_{pic}} \left[\sqrt{\frac{2E_x}{m_x^*}} \frac{q}{\pi\hbar} \sqrt{\frac{m_x^*}{2E_x}} \frac{1}{1+\exp\left(\frac{E_x+E_i+E_{kj}-\mu_S}{k_BT}\right)} \right] dE_x \; -$$

$$\int_{E_{pic}}^{\infty} \left[\sqrt{\frac{2E_x}{m_x^*}} \frac{q}{\pi\hbar} \sqrt{\frac{m_x^*}{2E_x}} \frac{1}{1+\exp((E_x+E_i+E_{kj}-\mu_D)/k_BT)} \right] dE_x \qquad [3.19]$$

where:

- $\sqrt{2E_x/m_x^*}$ represents the drift velocity;

- q is the elementary charge.

Due to opposite signs of electron velocity, the integral contribution disappears only from the third integral. Then, equation [3.19] becomes:

$$J(E_{kj}) = \frac{q}{\pi\hbar} \int_{E_{pic}}^{\infty} \left[\frac{1}{1+\exp((E_x+E_i+E_{kj}-\mu_S)/k_BT)} - \frac{1}{1+\exp((E_x+E_i+E_{kj}-\mu_D)/k_BT)} \right] dE_x \quad [3.20]$$

By integrating on the E_{kj}, we obtain:

$$J = \int_0^{\infty} \frac{1}{\pi\hbar} \sqrt{\frac{m_y^*}{2E_{kj}}} J(E_{kj}) dE_{kj} \qquad [3.21]$$

$$J = \frac{q}{\hbar^2} \sqrt{\frac{m_y^*}{2}} \left(\frac{-k_BT}{\pi}\right)^{3/2} \left[\mathfrak{I}_{1/2}(\tilde{\mu}_S - \tilde{E}_{pic}) - \mathfrak{I}_{1/2}(\tilde{\mu}_D - \tilde{E}_{pic}) \right] \qquad [3.22]$$

3.3.4. *Numerical methods*

Reducing the dimensions of devices to nanoscale sizes reveals the quantum nature of the phenomena. This affects transportation significantly.

To determine the electrical characteristics of this device, it is necessary to establish a mathematical model that describes the electronic transport considering the quantum phenomena that exist.

Different formalisms exist; here, we present the one that seems the most appropriate.

3.3.4.1. *Wigner function*

At the nanoscale, the wave nature of electrons takes its full force and the approach of the semi-classical Boltzmann transport, usually used to model the behavior of devices, becomes insufficient. The extension of this formalism to the quantum case leads to the Wigner transport equation, which bears a strong resemblance to that of Boltzmann.

The Wigner equation is derived from the Schrödinger equation and the density matrix function (DMF). It consists of combining the Schrödinger equation and the DMF by applying a Fourier transform.

Indeed, the Schrödinger equation defines the wave function of an electron that moves in a given profile potential and a certain energy. The DMF is an operator that correlates the wave function at different points in the real space and the space of the wave vectors via a Fermi-Dirac distribution.

Therefore, the solution of the Wigner function corresponds to a distribution of the wave functions through these vector spaces.

3.3.4.2. *Formalism*

A single particle of mass m is considered to move with a potential energy $V(x)$. The phase space is defined by the Cartesian product of position x and momentum p. The physical quantities are dynamic functions $A(x, p)$ of the coordinates of the phase space, such as kinetic and potential energies and their sum giving the Hamiltonian $H(x, p)$. The state of the particle at a given moment is presented by a point in phase space. Provided that the initial coordinates of the particles are known, the pair of coordinates $x(t)$, $p(t)$ at the instant t is obtained from Hamilton equations:

$$\dot{x} = \frac{\partial H(x,p)}{\partial p} = \frac{p}{m} \tag{3.23}$$

$$\dot{p} = \frac{\partial H(x,p)}{\partial x} = -\frac{\partial V(x)}{\partial x} \tag{3.24}$$

The function A(t) describes how physical quantities change over time. Two paths are possible: (a) $A(t) = A(x(t), p(t))$ is the old function of the new coordinates; (b) $A(t) = A(t, x, p)$ is a new function of the old coordinates. In the first case, we postulate that the laws of mechanics do not change with time: A remains the same function for the old and new coordinates. Then, with the help of equations [3.23] [3.24], we obtain the equation of evolution of A.

$$\dot{A} = \frac{\partial A(x,p)}{\partial x}\frac{\partial H(x,p)}{\partial p} - \frac{\partial A(x,p)}{\partial p}\frac{\partial H(x,p)}{\partial x} = [A, H]_P, [x, p]_P = 1 \qquad [3.25]$$

A fundamental notion regarding dynamic functions is that of the Poisson bracket $[\bullet, \bullet]$ P. It gives rise to an automorphism (preserving the algebraic structure) for mapping of all these functions.

In the second case, we must postulate a law of evolution of $A(t, x, p)$. If it is imposed according to equation [3.25], automorphism systematically leads to the conservation of the laws of mechanics: the new function in old coordinates is the old function in new coordinates: $A(t, x, p) = A(x(t), p(t))$.

A statistical description is introduced if the coordinates of the point cannot be declared exactly, but with some probability. According to the basic premise of classic statistical mechanics, the state of the particle system is completely defined by a function $f(x, p)$, with the following properties:

$$f(x,p) \geq 0; \quad \int dx.\, dp.\, f(x, p) = 1 \qquad [3.26]$$

The physical quantities are described by the corresponding mean values:

$$\langle A \rangle(t) = \int dx.\, dp.\, A(t, x, p).\, f(x, p) \qquad [3.27]$$

This equation is impractical because it requires the calculation of the evolution of a particular quantity A. However, due to the automorphism of the Poisson bracket, variables can be modified such that time is transferred to the distribution function f. Equation [3.27] becomes:

$$\langle A \rangle(t) = \int dx dp A(x, p) f(x, p, t) \qquad [3.28]$$

The equation of the evolution of f can then be derived:

$$\left(\frac{\partial}{\partial t} + \frac{p}{m}.\frac{\partial}{\partial x} + F(x)\frac{\partial}{\partial p}\right) f(x, p, t) = \left(\frac{\partial f}{\partial t}\right)_c \qquad [3.29]$$

Here, force $F = -\nabla_x V$ is given by the derivative of potential energy V. The characteristics of the differential operator in the brackets, called the Liouville operator, are conventional Newtonian trajectories. The left-side trajectories of equation [3.29] become a total derivative. In the case of no interaction with the environment, $\partial f/\partial t_c = 0$, that is, trajectories have a constant value. Otherwise, the particles are redistributed between the trajectories and the right side of equation [3.29] is equal to the net change in particle density due to collisions.

3.3.4.3. Quantum operators

We quickly recall the principles of quantum mechanics of operators, which will be reformulated in the formalism of phase space. The physical quantities in quantum mechanics are presented by the Hermitian operators:

$$\hat{A}|\phi_n\rangle = a_n|\phi_n\rangle; \quad \langle\phi_n|\phi_m\rangle = \delta_{mn} \quad \sum_n |\phi_n\rangle\langle\phi_n| = \hat{1} \qquad [3.30]$$

These operators have real eigenvalues and a complete, orthonormal system of eigenvectors that form a Hilbert space. The states of the system are specified by elements $|\psi\_t>$ of the Hilbert space (integrable squares, i.e. a finite power) and normalized with respect to the L_2 standard in H. In wave mechanics, the evolution of $|\psi\_t>$ is postulated to be provided by the Schrödinger equation:

$$\hat{H}|\Psi_t\rangle = i\hbar\frac{\partial|\Psi_t\rangle}{\partial t} \quad \langle\Psi_t|\Psi_t\rangle = 1 \quad |\Psi_t\rangle = \sum_n c_n(t)|\phi_n\rangle \qquad [3.31]$$

The state can be broken down into a complete base of an observable A. In addition, it can be shown that, during evolution, the state remains normalized. This property is often called probability conservation.

According to the principle of correspondence, conventional position and momentum variables correspond the Hermitian operators x and p, satisfying a quantum counterpart of the Poisson bracket:

$$x \rightarrow \hat{x} \quad p \rightarrow \hat{p} \quad \hat{x}\hat{p} - \hat{p}\hat{x} = [\hat{x}, \hat{p}]_- = i\hbar\hat{1} \qquad [3.32]$$

Wave mechanics uses only half of the phase space-representation in coordinates or momentum for the description of the physical system. We use a representation in coordinates:

$$\hat{x}|x\rangle = x|x\rangle \quad \int dx|x\rangle\langle x| = \hat{1} \quad \hat{p} = -i\hbar\frac{\partial}{\partial x} \qquad [3.33]$$

Finally, we recall the equation for the mean value of a physical quantity:

$$\langle A \rangle(t) = \langle \Psi_t | \hat{A} | \Psi_t \rangle = \int dx \langle \Psi_t | x \rangle \langle x | \hat{A} | \Psi_t \rangle \qquad [3.34]$$

The operator's formulation of quantum mechanics seems too abstract when compared to familiar classic concepts. Nevertheless, it is possible to reformulate the ideas of quantum mechanics in phase space.

We have:

$$\langle x | \hat{A} | \Psi_t \rangle = \int dx' \sum_n a_n \langle x | \phi_n \rangle \langle \phi_n | x' \rangle \langle x' | \Psi_t \rangle = \int dx' \alpha(x, x') \Psi_t(x') \qquad [3.35]$$

where: $\psi_t(x) = \langle x | \psi_t \rangle$. A substitution in [3.35] shows that the physical mean is actually evaluated in a "double half" of the phase space:

$$\langle A \rangle(t) = \int dx' \int dx \, \alpha(x, x') \rho_t(x', x) = Tr(\hat{\rho}_t \hat{A}) \qquad [3.36]$$

With ρ_t and $\hat{\rho}_t$, the density matrix and density operator are given as:

$$\rho_t(x, x') = \Psi_t^*(x') \Psi_t(x) = \langle x | \Psi_t \rangle \langle \Psi_t | x' \rangle = \langle x | \hat{\rho}_t | x' \rangle \qquad \hat{\rho}_t = \sum_{m,n} c_m^*(t) c_n(t) | \phi_n \rangle \langle \phi_m |$$

$$[3.37]$$

3.3.4.4. Wigner function for pure state

We then get the Von Neumann equation of motion for the pure state density matrix ρ_t:

$$i\hbar \frac{\partial \rho(x, x', t)}{\partial t} = \langle x | [\hat{H}, \hat{\rho}_t] - | x' \rangle =$$
$$\left\{ -\frac{\hbar^2}{2m} \left(\frac{\partial^2}{\partial x^2} - \frac{\partial^2}{\partial x'^2} \right) + (V(x) - V(x')) \right\} \rho(x, x', t) \qquad [3.38]$$

Variables are modified using a center of mass transformation:

$$x_1 = \frac{(x + x')}{2}; \quad x_2 = x - x' \qquad [3.39]$$

$$\frac{\partial \rho \left(x_1 + \frac{x_2}{2}, x_1 - \frac{x_2}{2}, t \right)}{\partial t} = \frac{1}{i\hbar}$$
$$\left\{ -\frac{\hbar^2}{m} \frac{\partial^2}{\partial x_1 \partial x_2} + (V(x_1 + x_2/2) - V(x_1 - x_2/2)) \right\} \rho \left(x_1 + \frac{x_2}{2}, x_1 - \frac{x_2}{2}, t \right) \qquad [3.40]$$

The Wigner function is obtained by using the Fourier transform with respect to x_2:

$$f_w(x_1, p, t) = \frac{1}{(2\pi\hbar)} \int dx_2 \rho(x_1 + x_2/2, x_1 - x_2/2, t) e^{-ix_2 \cdot p/\hbar}$$

[3.41]

We note that, due to the Wigner transform, x_1 and p are independent variables. It is easy to show that the corresponding operators switch. Thus, x_1 and p define a phase space – the Wigner phase space. The Fourier transform of the right-hand side of [3.41] gives rise to two terms, which are evaluated as follows. It is convenient to introduce the abbreviation:

$$\rho(+, -, t) \text{ for } \rho\left(x_1 + \frac{x_2}{2}, x_1 - \frac{x_2}{2}, t\right):$$

$$I = -\frac{1}{i\hbar}\frac{\hbar^2}{m(2\pi\hbar)} \int dx_2 e^{-ix_2 \cdot \frac{p}{\hbar}} \frac{\partial^2 \rho(+, -, t)}{\partial x_1 \partial x_2} = -\frac{1}{m} p \cdot \frac{\partial f_w(x_1, p, t)}{\partial x_1}$$

[3.42]

where we integrated by parts and used the fact that the density matrix tends toward zero at infinity: $\rho \to 0$ si $x_2 \to \pm\infty$ 0 $\to$

$$II = \frac{1}{i\hbar(2\pi\hbar)} \int dx_2 e^{-ix_2 \cdot \frac{p}{\hbar}} (V(x_1 + x_2/2) - V(x_1 - x_2/2)) \rho(+, -, t) =$$

$$\frac{1}{i\hbar(2\pi\hbar)} \int dx'e^{-ix_2 \cdot \frac{p}{\hbar}} \left(V\left(x_1 + \frac{x_2}{2}\right) - V\left(x_1 - \frac{x_2}{2}\right)\right) \cdot \delta(x_2 - x')\rho\left(x_1 + \frac{x'}{2}, x_1 - x'/2\right) \text{ [3.43]}$$

After replacing the delta function with the following integral:

$$\delta(x_2 - x') \text{ by } \frac{1}{(2\pi\hbar)} \int dp' e^{i(x_2 - x')p'/\hbar}$$

[3.44]

it so happens that:

$$II = \frac{1}{i\hbar(2\pi\hbar)}$$

$$\int dp' \int dx_2 e^{-ix_2 \cdot \frac{p-p'}{\hbar}} (V(x_1 + x_2/2) -$$

$$V(x_1 - x_2/2)) \cdot \frac{1}{(2\pi\hbar)} \int dx'e^{-ix' \cdot \frac{p'}{\hbar}} \rho\left(x_1 + \frac{x'}{2}, x_1 - \frac{x'}{2}; t\right) =$$

$$\int dp' V_w(x_1, p - p') f_w(x_1, p') t$$

[3.45]–[3.46]

We summarize the results of these transformations. This leads to Wigner equation:

$$\frac{\partial f_w(x, p, t)}{\partial t} + \frac{p}{m} \cdot \frac{\partial f_w(x, p, t)}{\partial x} = \int dp' V_w(x, p - p') f_w(x, p', t)$$

[3.47]

where V_w is the Wigner potential.

$$V_w(x,p) = \frac{1}{i\hbar(2\pi\hbar)} \int dx'\, e^{-ix'\frac{p}{\hbar}}\left(V(x+x'/2) - V(x-x'/2)\right) \qquad [3.48]$$

A change in the sign of x' reveals the antisymmetry of the Wigner potential.

3.3.4.5. *Classic limit of the Wigner equation*

We discuss the classic limit of [3.48] considering the case where the potential Vi is a linear or quadratic function of the position, namely:

$$V\left(x \pm \frac{x'}{2}\right) = V(x) \pm \frac{\partial V(x)}{\partial x}\frac{x'}{2} + \cdots = V(x) \mp F(x)\frac{x'}{2} + \cdots \qquad [3.49]$$

The dotted lines represent the quadratic term. Force F can be at most a linear function of the position. As the even terms of the Taylor series of V cancel out in [3.48], the Wigner potential becomes:

$$V_w(x,p) = \frac{1}{\hbar(2\pi\hbar)} \int dx'\, e^{-ix'\frac{p}{\hbar}} F(x)x' \qquad [3.50]$$

The right hand side of [3.47] becomes:

$$\int dp'\, V_w(x, p-p')f_w(x,p',t) =$$
$$\frac{i}{\hbar(2\pi\hbar)} \int dp' \int dx'\, e^{-ix'\frac{p-p'}{\hbar}} F(x)x' f_w(x,p',t) =$$
$$\frac{-F(x)}{2\pi\hbar}\frac{\partial}{\partial p} \int dp' \int dx'\, e^{-ix'\frac{p-p'}{\hbar}} f_w(x,p',t) = -F(x)\frac{f_w(x,p,t)}{\partial p} \qquad [3.51]$$

where we used equality:

$$ix'e^{-ix'\frac{p-p'}{\hbar}} = -\hbar\frac{\partial}{\partial p}e^{-ix'\frac{p-p'}{\hbar}} \qquad [3.52]$$

Then, the Wigner equation is reduced to the Boltzmann equation, without collision:

$$\frac{\partial f_w(x,p,t)}{\partial t} + \frac{p}{m}\cdot\frac{\partial f_w(x,p,t)}{\partial x} + F(x)\frac{\partial f_w(x,p,t)}{\partial p} = 0 \qquad [3.53]$$

Now, let us consider as an initial condition a minimum uncertainty on the wave packet. The Wigner function of such a packet is a Gaussian of the position and momentum. The latter can equally be interpreted as a conventional initial distribution of electrons, provided that force is a constant or linear function of the position; the packet evolves like that of the conventional distribution. Despite the spreading in the phase space, Gaussian devices determine the general shape of the packet. f_w remains positive during the evolution. However, strong variations in the field with position introduce interference effects. Near band shifts, the packet quickly loses its shape and negative values appear.

Monte Carlo algorithms can be designed based on the idea that conditions on the right-hand side of the Wigner–Boltzmann equation represent gain and loss of terms for the density of the phase space. To fix the ideas, consider the semi-classic Boltzmann equation.

$$\left(\tfrac{\partial}{\partial t} + v(k).\nabla_r + \tfrac{1}{h}F(r).\nabla_k\right)f(r,k,t) = \int dk' f(r,k',t)S(k',k) - \lambda(k)f(r,k,t) \quad [3.54]$$

3.3.4.6. *On the Boltzmann equation*

We seek to solve PDEs, starting from complex models, to simplify them in the case of certain types of regimes. Indeed, applying the physical laws such as mechanics to all particles, considering positions and velocities, is prohibitive: there are too many degrees of freedom and too many equations.

Simplifications make it possible to consider only the velocities, and to arrive at the Navier–Stokes equations, for example, or in any case at some of its reductions. Mathematicians have been able to arrive at the equation of heat, via the linear Boltzmann equation, itself a master equation for the kinetic theory of collisions.

Some hypotheses:

– particles are indistinguishable;

– the duration of collisions is zero (hard spheres, elastic shocks);

– a writing of the Boltzmann equation:

$$\frac{\partial f_v}{\partial t} + V_v(k).\frac{\partial f_v}{\partial x} - \frac{q}{h}E(x,t).\frac{\partial f_v}{\partial k} = Q_v(f_v) + \sum_{\mu \neq v} Q_{\mu \to v}(f_v, f_\mu) = \left(\frac{\partial f_v}{\partial t}\right)_{coll.} \quad [3.55]$$

– considering a gas consisting of n particles, with $10^{22} > n > 10^6$.

A particle can already represent a statistical average of a certain number of, for example, electric carriers. The microscopic study of the Boltzmann function, representing a carrier probability density, is carried out in phase space: three dimensions of positions x, y, z, and three velocities v_x, v_y, v_z or of wave vector (m^{-1}) – k_x, k_y k_z (in the case of a parabolic energy band: $\varepsilon e = \hbar^2 k^2 / 2m$, $v = \hbar k/m$). The set of particles can be likened to a gas "rarefied enough" to assume that at least three particles cannot interact simultaneously.

The second term on the left, transport $\vec{v}.\vec{\nabla}_x.f$, indicates that particles propagate in a straight line.

Collision term Q (scattering) is the input term–output term balance on a wave vector state $k(m^{-1})$.

NOTE.– During numerical calculations, it may be useful to keep the output term of Q in integral form, and not to actually calculate its integral, with numerical errors on the two input/output terms being subtracted, tending to compensate for each other.

Here, we will only mention that it is possible to solve the so-called Schur equations, a simplification of the Boltzmann model, but we still need to know the stationary regime.

The fluid is assumed to be incompressible (in this case, its density is constant: this results in a zero divergence of the velocity field of all the particles considered), occupying the same volume before and after the shocks; this is not really compatible when coupled with the Poisson equation, which translates the attraction or repulsion between carriers in a volume element.

In fact, a kinetic approach is much more complex than a hydrodynamic description, because it works in phase space; the input term of the collision integral is also "expensive" in calculation time.

Moving from Boltzmann kinetics to hydrodynamic models (asymptotic?) presupposes a thermodynamic equilibrium, even locally; in practice, the Boltzmann distribution f is invariable by the collision operator (on average, the input term is equal to the output term, over a state k, which implies a Maxwellian form of f, associated with a maximum entropy (see entropic distance), locally (see Gibbs lemma)). For small disturbances, at small electric fields, for example, the solutions of f can still be considered as Maxwellian, but displaced along this field.

Moreover, can we deduce fluid equations from Newtonian mechanics?

– Hilbert (David Hilbert, 1862 – Königsberg, 1943 – Göttingen; "*Wir müssen wissen, wir werden wissen*", that is, "We must know, we will know") proposed to use the Boltzmann equation; this would de facto restrict the range of possible solutions, particularly because of the introduction of the ideal gas model.

The Boltzmann equation could also be considered by an approximation of that of compressible Euler equations, and this for small Kundsen numbers (linked to the ratio of the mean free path to the nominal size of the system considered).

From the Boltzmann equation, it is also possible to deduce certain forms of the Navier–Stokes equation (for certain temperature laws).

– There is a lack of knowledge of compressible equations.

The compressible Navier–Stokes equation cannot be deduced a priori from the Boltzmann equation, but it could be deduced from the compressible Euler equation, which is another master equation.

If the fluctuation of f around the equilibrium, as for the Navier–Stokes equation, seems very difficult to understand, what about the modeling of shock waves or chaos?

In any case, many mathematical works, at the highest level, are in full effervescence, especially with regard to asymptotic developments of such equations.

The problems related to "device modeling" are a priori less complex for solving parabolic diffusion equations than for the nonlinear elliptical Poisson equation, or for low applied voltage levels, where the quasi-Fermi level is often used. The three equations, "Poisson" and divergence of electron and hole current densities, are then linear elliptical.

On the other hand, if we define, for example at the contacts, Schottky conditions, in exponential form of $\exp(qV/kT)$, then equations linked to current density divergence become linear elliptical; "underflows" or "overflows" may appear, linked to these exponentials.

On the other hand, at a high voltage or electric field level, a strong coupling prevails between the three equations. The terms of current drift due to the electric field become predominant, as boundary layers on the quasi-Fermi levels are able to induce numerical difficulties. As for the end of the displacement current, which is all the higher as the frequency is higher, it is sometimes preferable to keep it, even at low frequency, for numerical stability problems. Each equation is linear in the main variable.

If recombination generation terms are not introduced, equations relating to electrons and holes lead to a non-symmetrical system, with the resolution being more complex than for symmetrical systems.

Based on the Boltzmann equation, several models have been derived. Statistical methods such as those of Monte Carlo seem to be the most suitable for the simulation of submicron devices; moreover, the smaller the devices, the less time-consuming they could be, the limit being a particle <–> a carrier.

When we enter the strongly submicron domain, Monte Carlo methods seem the most interesting, coupled moreover with a Coulomb's law, to take into account the interaction between carriers, thus abandoning the Poisson equation; at a given mesh, we can then calculate distances between barycenters of meshes taken 2 to 2, and store them at the beginning of the program; then, it suffices to count incoming particles and count outgoing ones to readjust the electric field in the meshes considered.

The velocity auto-correlation function, which is easy to introduce into these statistical algorithms, should be the basic tool for the analysis of the scattering noise by taking the collision times of carriers with phonons of the network into account. Also, the duration of shocks and the exchange of phonons for inelastic interactions will have to consider the Heisenberg paradigm (eDt > h/2/p).

If (k',k) denotes the rate from the initial state k' to the final state k, induced by the transition physical diffusion processes, and λ is the total diffusion rate, we have:

$$\lambda(k) = \int dk' S(k, k') \tag{3.56}$$

In a Monte Carlo algorithm, the term λ_f gives rise to the exponential distribution for free flight time of the carrier.

The Wigner–Boltzmann equation then has the following structure, added by a Boltzmann diffusion operator:

$$\left(\frac{\partial}{\partial t} + v(k).\nabla_r + \frac{1}{h}F_{cl}(r).\nabla_k\right) f_w(r, k, t) = \int dk' \Gamma(k, k')\mu(k')f_w(r, k', t) - \mu(k)f_w(r, k, t) \tag{3.57}$$

The integral kernel Γ, in this equation, has the form

$$\Gamma(r, k, k') = \frac{1}{\mu(r, k')}[S(k', k) + V_m(r, k - k') + \alpha(k, r)\delta(k - k')] \tag{3.58}$$

$$\mu(r, k') = \lambda(r, k') + \alpha(r, k') \tag{3.59}$$

where μ is the normalization factor. We have:

$$\int dk' \Gamma(k, k', r) = 1 \tag{3.60}$$

We can introduce a fictitious collision mechanism:

$$S_{self}(k', k) = \alpha(r, k)\delta(k - k'), \tag{3.61}$$

which is called auto-diffusion. Mathematically, the related contributions in the terms of gain and loss are simply cancelled and have no effect physically; due to the

δ function, this mechanism does not change the state of the electron and therefore does not modify the free flight trajectory. The choice of α offers a degree of freedom in the construction of a Monte Carlo algorithm.

Finally, we can conclude that the Monte Carlo methods seem well suited to the study of these devices (DGMOS) compared to conventional numerical methods, often unstable for the dynamic regime.

3.3.4.7. *Integral form of the Wigner–Boltzmann equation*

The Wigner–Boltzmann equation can be transformed into an integral path equation: the added integral equation, which will give rise to the use of Monte Carlo algorithms, via the following integral kernel:

$$P(k_f, t_f | k_i, t_i) = \Gamma[k_f, K(t_f)]\mu[K(t_f)]\exp\left\{-\int_{t_i}^{t_f}\mu[K(\tau)]d\tau\right\} \qquad [3.62]$$

The kernel represents a transition composed of a free flight from the time t_i with initial state k_i, followed by a diffusion process at the final state k_f at the time t_f. For the sake of brevity, the dependencies of Γ and μ are omitted below. In a Monte Carlo simulation, the duration of the next diffusion event, T_f, is generated from the exponential distribution appearing in [3.62]:

$$p_t(t_f, t_i, k_i) = \mu[K(t_f)]\exp\left\{-\int_{t_i}^{t_f}\mu[K(\tau)]d\tau\right\} \qquad [3.63]$$

The state at the end of free flight is denoted k' = $K(t_f)$. A transition from the final trajectory point k' to the final state k_f is carried out using the kernel Γ. Unlike the classic case, where P would represent a transition probability, such an interpretation is not possible in the case of the Wigner equation, because P is not semi-defined positive. The problem comes from the "Wigner" potential, which takes positive and negative values. Due to its anti-symmetry properties with respect to q, the Wigner potential can be reformulated in terms of a positive function V^+_w:

$$V^+_w(r, q) = \max\left(0, V_w(r, q)\right) \qquad [3.64]$$

$$V_w(r, q) = V^+_w(r, q) - V^+_w(r, -q) \qquad [3.65]$$

Next, the kernel Γ is rewritten as a sum of the following conditional probability distributions:

$$\Gamma(k, k') = \frac{\lambda}{\mu}\, s(k', k) + \frac{\alpha}{\mu}\delta(k' - k) + \frac{\gamma}{\mu}[w(k, k') - w^*(k, k')] \qquad [3.66]$$

$$s(k', k) = \frac{S(k', k)}{\lambda(k')}, \quad w(k, k') = \frac{V^+_w(k - k')}{\gamma}, \quad w^*(k, k') = w(k', k) \qquad [3.67]$$

The normalization factor of the Wigner potential is:

$$\gamma(r) = \int dq V^+(r, q) \qquad [3.68]$$

3.3.4.8. Schrodinger–Poisson model

The model considered for establishing the electric characteristic of the device is therefore based on Poisson and Schrödinger equations. Coupling of the Poisson and Schrödinger equations is necessary when the region of the oxide is decreasing and the region of the channel takes values close to the wavelength of electrons. This approach is justified by several results presented in the appropriate literature.

3.3.4.9. Schrodinger–Poisson model (recap)

The integration of the Poisson equation makes it possible to calculate the potential variation in the semiconductor. The space charge is formed by charges due to carriers and free impurities, which are assumed to be completely ionized.

The Poisson equation in two dimensions is written (recap) as follows:

$$\frac{d^2V(x,y)}{dx^2} + \frac{d^2V(x,y)}{dy^2} = -\frac{q}{\varepsilon}[p - n + N_D - N_A] \qquad [3.69]$$

where:

– V: electric potential;

– p: concentration of holes;

– N: electron concentration;

– ND and NA: concentrations of donors and acceptors;

– q: elementary charge;

– ε: dielectric constant.

3.3.4.10. Schrödinger equation

If the potential variations are known in advance, energy levels and wave functions in the quantum confinement direction (along the y-axis) can be obtained by solving the Schrödinger equation.

We get the Schrödinger equation from:

$$-\frac{\hbar^2}{2m^*}\frac{\partial^2}{\partial y^2}\psi(y) + qV(y)\psi(y) = E\psi(y) \qquad [3.70]$$

where:

- ψ (y): wave function;

- E: quantum energy;

- h: Planck's constant;

- V(y): electrostatic potential.

We can illustrate the self-consistent system by the following system:

$$\begin{cases} \rho(y) = S[V(y)] \\ V(y) = P[\rho(y)] \end{cases}$$

[3.71]

where functions S[V(y)] and P[ρ(y)] represent the Poisson and Schrödinger equations, respectively.

3.3.4.11. *Discretization methods*

There are several discretization methods for solving the coupled equations considered ("Poisson–Schrodinger"). The most widely used are finite difference method (FDM) and finite element method (FEM). For now, we have adopted the finite difference method using a constant rectangular mesh.

The choice of this method is guided by:

- the ease of implementation;

- the numerical stability to solve these coupled equations.

The general equation to be discretized has the following form:

$$\frac{\partial}{\partial x}\left[P(x,y).\frac{\partial U}{\partial x}\right] + \frac{\partial}{\partial y}\left[P(x,y).\frac{\partial U}{\partial y}\right] = f(U,x,y)$$

[3.72]

where:

- U: unknown function;

- P ef: functions determined a priori.

Considering the integration of equation [3.72] and a Green transformation, we can discretize:

$$G_k \cdot U_{k-1} + B_k \cdot U_{k-n} + D_k \cdot U_{k+1} + H_k \cdot U_{k+1} - C_k \cdot U_k = f(U_k, x_i, y_j) \qquad [3.73]$$

where:

- i: index along the x axis;

- j: index along the y axis;

- k: global index defined by: $k = (j - i) * n + 1$;

- $D_k = P\left(\frac{x_i + x_{i+1}}{2}, y_j\right) \cdot \frac{2}{(x_{i+1} - x_i) \cdot (x_{i+1} - x_{i-1})}$;

- $G_k = P\left(\frac{x_i + x_{i-1}}{2}, y_j\right) \cdot \frac{2}{(x_i - x_{i-1}) \cdot (x_{i+1} - x_{i-1})}$;

- $B_k = P\left(x_i, \frac{y_j + y_{j-1}}{2}\right) \cdot \frac{2}{(y_j - y_{j-1}) \cdot (y_{j+1} - y_{j-1})}$;

- $H_k = P\left(x_i, \frac{y_j + y_{j+1}}{2}\right) \cdot \frac{2}{(y_{j+1} - y_j) \cdot (y_{j+1} - y_{j-1})}$;   [3.74]

- $C_k = G_k + D_k + B_k + H_k.$

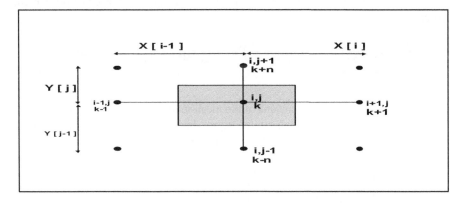

Figure 3.10. *Mesh cell*

3.3.4.11.1. Solving the Poisson equation

3.3.4.11.1.1. MDF discretization

Figure 3.11 shows an example of a mesh of a DGMOS transistor.

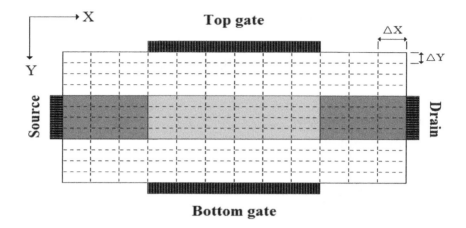

Figure 3.11. *Example of a mesh of a double-gate MOS transistor*

The domain of the calculation is (NX × NY) nodes, where NX and NY are the number of nodes in the x and y directions, respectively. The numerical solution of the Poisson equation is composed of (NX x NY) potential values.

Applying the finite difference method (FDM) to the Poisson equation, we can write:

$$\frac{\Delta y}{\Delta x}V_{i-1,j} + \frac{\Delta x}{\Delta y}V_{i,j-1} - 2\left(\frac{\Delta x}{\Delta y} + \frac{\Delta x}{\Delta y}\right)V_{i,j} + \frac{\Delta x}{\Delta y}V_{i,j+1} + \frac{\Delta y}{\Delta x}V_{i+1,j} = -\Delta x.\,\Delta y - \frac{q}{\varepsilon}[n + N_D - N_A]$$

$$[3.75]$$

where Dx and Dy are the dimensions of the mesh in dimension x and y, respectively. It is possible to write the Poisson equation discretized as the matrix formula: S = M.X, where M is a matrix with G, B, D, H and C: coefficients:

$$[M] = \begin{bmatrix} C & D & . & . & . & H & . & . & . \\ G & C & D & . & . & . & H & . & . \\ . & G & C & D & . & . & . & H & . \\ . & . & G & C & D & . & . & . & H \\ . & . & . & G & C & D & . & . & . \\ B & . & . & . & G & C & D & . & . \\ . & B & . & . & . & G & C & D & . \\ . & . & B & . & . & . & G & C & D \\ . & . & . & B & . & . & . & G & D \end{bmatrix}$$

$$[3.76]$$

With coefficients: $G_k = D_k = \frac{\Delta y}{\Delta x}$, $B_k = H_k = \frac{\Delta x}{\Delta y}$

In the case where nodes are positioned at the silicon/oxide interface, the discontinuity permittivity e must be considered. Indeed, there is continuity of the displacement vector if there are no interfacial charges.

$$\frac{\Delta y}{\Delta x} V_{i-1,j} + \frac{\Delta x}{2\Delta y}\left(1 + \frac{\varepsilon_{bot}}{\varepsilon_{top}}\right) V_{i,j-1} - \left(\frac{\Delta x}{\Delta y} + \frac{\Delta x}{\Delta y}\right)\left(1 + \frac{\varepsilon_{bot}}{\varepsilon_{top}}\right) V_{i,j} + \frac{\Delta x}{2\Delta y} \qquad [3.77]$$

$$\left(1 + \frac{\varepsilon_{bot}}{\varepsilon_{top}}\right) V_{i,j+1} + \frac{\Delta y}{\Delta x}\frac{\varepsilon_{bot}}{\varepsilon_{top}} V_{i+1,j} = -\Delta x.\Delta y - \frac{q}{\varepsilon_{top}}[n + N_D - N_A]$$

where ε_{bot} and ε_{top} are the dielectric constants for the material above and below the interface. In the case considered, the two oxides are identical.

3.3.4.11.1.2. Conventional boundary and contact conditions

To solve these equations, boundary conditions must be considered; they are of the Dirichlet or Neumann type.

Gate contacts correspond to a metallic material deposited on an oxide gate material; then the Dirichlet boundary conditions are to be applied, which means that the potential V must be as follows:

$$V = V_G \qquad [3.78]$$

Gate potential is determined from the gate bias voltage V_G. This corresponds to the following numerical equation:

$$V_{i,j} = V_g \qquad [3.79]$$

Indeed, drain and source contacts correspond to a metallic material deposited on the semiconductor substrate, so that the Dirichlet boundary condition is then written as:

$$V_{i,j} = V_D + U_T \, Ln \, \frac{N_D}{n_i} \qquad [3.80]$$

$$V_{i,j} = V_S + U_T \, Ln \, \frac{N_D}{n_i} \qquad [3.81]$$

3.3.4.11.1.3. Other boundary conditions

Neumann conditions are applied on the other boundary surfaces of the semiconductor. These boundary conditions are necessary to ensure the neutrality of charges on contact zones:

$$\begin{cases} \frac{\partial V}{\partial y} = 0, \frac{\partial n}{\partial y} = 0 \end{cases} \qquad [3.82]$$

The Neumann boundary conditions are as follows:

– for the left and right edges:

$$Vi, j - Vi \pm 1, j = 0; \qquad [3.83]$$

– for the upper and lower edges:

$$Vi, j - Vi, j \pm 1 = 0; \qquad [3.84]$$

– for the two corner nodes along the top edge:

$$2Vi, j - Vi + 1, j + Vi, j \pm 1 = 0; \qquad [3.85]$$

– for the two corner nodes along the lower edge:

$$2Vi, j - Vi + 1, j + Vi, j \pm 1 = 0. \qquad [3.86]$$

So far, we have obtained the (NX × NY) equations necessary to solve the Poisson equation and determine the potential Vi, j at each node of the mesh.

Considering ND, NA and n, the preceding equations represent a set of linear equations which can be solved directly by considering an iterative method. However, when solving a coupled set of equations, there is a better solution algorithm for solving the Poisson equation. This algorithm consists of a change of variable for n, namely expressing it in terms of the Fermi potential and quasi-energy level, Fn. The potential energy of the quasi-Fermi level is calculated as a function of the old potential:

$$(F_n)_{i,j} = -q(V_{old})_{i,j} + k_B T . \mathfrak{I}_{1/2}^{-1} \left(\frac{n_{i,j}}{N_C} \right) \qquad [3.87]$$

where:

– $\mathfrak{I}_{1/2}^{-1}$ is the Fermi integer of order ½;

– N_C is the density of states in the conduction band. The electron density is then:

$$n_{i,j} = N_C \mathfrak{I}_{1/2}\left[\frac{(F_n)_{i,j} + qV_{i,j}}{k_B T}\right]. \tag{3.88}$$

With this change in variable, basic equations represent a set of nonlinear equations for the potential. The Poisson equation is then solved by the Newton–Raphson method such that:

$$F\left(\vec{V}_i\right) = 0 \tag{3.89}$$

where $F\left(\vec{V}_i\right)$ is the Poisson equation function. The Jacobian matrix is obtained by:

$$F_{\alpha,\beta}(V) \equiv \frac{\partial F_\alpha(V)}{\partial V_\beta} \tag{3.90}$$

Given an initial assumption or an old V_{old} solution, the projected solution is $V_{new} = V_{old} + \Delta V$.

Using a Taylor series development of the first order, we can obtain the new potential value such that:

$$F_\alpha(V_{new}) \approx F_\alpha(V_{old}) + F_{\alpha,\beta}(V_{old}).[\Delta V]_\beta = 0 \tag{3.91}$$

It follows that updates can be obtained as:

$$[\Delta V]_\beta = -\frac{F_\alpha(V_{old})}{F_{\alpha,\beta}(V_{old})} \tag{3.92}$$

3.3.4.11.1.4. Solving the Schrödinger equation

In the DGMOS studied, it can be considered that the confinement is efficient in only one direction; therefore, the Schrödinger equation is then solved in one dimension.

To solve the Schrödinger equation, the structure is divided into slices in the transport direction. These are called Schrödinger slices. Figure 3.12 illustrates the double gate structure divided into Schrödinger slices.

In each slice (or column), the potential profile is obtained by solving the Poisson equation as before; consequently, solving the Schrödinger equation in the confinement direction (y axis) makes it possible to obtain energy levels and wave functions.

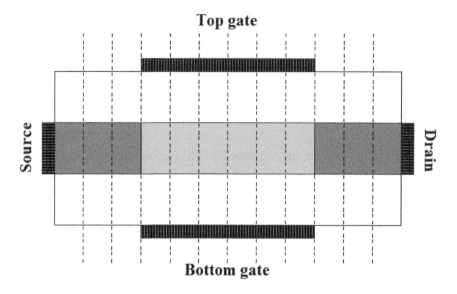

Figure 3.12. *Diagram of a DGMOS structure*
divided into "Schrödinger" slices

The Schrödinger equation in the confinement direction is written as follows:

$$-\frac{\bar{h}^2}{2m^*}\frac{\partial^2}{\partial y^2}\psi_i(y) + qV(y)\psi_i(y) = E_i\psi_i(y) \qquad [3.93]$$

For each $y = y_{i,j}$, the equation discretized by the finite difference method becomes:

$$-\frac{\bar{h}^2}{2m^*(y)}\frac{\psi_i(y_{i+1})-2\psi_i(y_i)+\psi_i(y_{i-1})}{\Delta y^2} + qV(y_i)\psi_i(y_i) = E_i\psi_i(y_i) \qquad [3.94]$$

Equation [3.94] represents the Schrödinger equation discretized in 1D.

We repeat this operation for all indices $j(1 \text{ to } N)$.

$$j = 0;\ -\frac{\bar{h}^2}{2m^*(y)}\frac{\psi_i(y_1)-2\psi_i(y_0)+\psi_i(y_{-1})}{\Delta y^2} + qV(y_0)\psi_i(y_{i0}) = E_i\psi_i(y_0) \qquad [3.95a]$$

$$j = 1 \; ; - \frac{\hbar^2}{2m^*(y)} \frac{\psi_i(y_2) - 2\psi_i(y_1) + \psi_i(y_0)}{\Delta y^2} + qV(y_1)\psi_i(y_1) = E_i\psi_i(y_1) \qquad [3.95b]$$

$$j = N \; ; - \frac{\hbar^2}{2m^*(y)} \frac{\psi_i(y_{N+1}) - 2\psi_i(y_N) + \psi_i(y_{N-1})}{\Delta y^2} + qV(y_N)\psi_i(y_N) = E_i\psi_i(y_N) \quad [3.96]$$

Therefore, the discretized Schrödinger equation described in matrix form becomes:

$$H\psi_i(y) = E_i\psi_i(y) \qquad [3.97]$$

where H is the Hamiltonian system defined by:

$$H = -\frac{\hbar^2}{2m^*(y)} \begin{bmatrix} -2 & 1 & 0 & \cdots & \cdots & 0 \\ 1 & -2 & 1 & 0 & \cdots & 0 \\ 0 & 1 & -2 & 1 & \cdots & 0 \\ \cdots & \cdots & \cdots & \cdots & \cdots & \cdots \\ 0 & \cdots & 0 & 1 & -2 & 1 \\ 0 & \cdots & \cdots & 0 & 1 & -2 \end{bmatrix} + q \begin{bmatrix} V(y_1) & 0 & \cdots & \cdots & 0 \\ 0 & V(y_2) & 0 \cdots & \cdots & 0 \\ \cdots & & \cdots & \cdots & \cdots \\ 0 & \cdots & 0 V(y_i) & 0 & \cdots \\ \cdots & \cdots & \cdots & \cdots & \cdots \\ 0 & \cdots & \cdots & \cdots & V(y_N) \end{bmatrix}$$

$$[3.98]$$

To solve the Schrödinger equation, it is necessary to know the electrostatic potential V(x,y), and to solve the Poisson equation, it is necessary to know the carrier concentration or the charge density ρ(x,y). There is, therefore, a consistency between the solutions of 2D Poisson equations and the 1D Schrödinger equations. These systems of equations are solved numerically; a self-consistent Newton Raphson method is therefore adopted. We obtained numerical results with precision of the order of 10^{-9}.

3.3.4.11.2. Numerical results

We first compare the electric characteristics obtained by classic and quantum models. It should be noted that the current level in quantum structures is lower than that observed on classic structures. Figure 3.13. shows the drain currents for Vds = 0.6V, showing the difference between the two models.

This is essentially due to the difference that exists with respect to the density of carriers present in the active zone.

If the dimensions of the transistor are of the order of a few tens of nanometers, the known laws of classic physics are replaced by those of quantum physics.

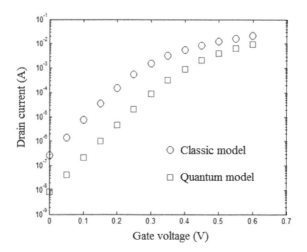

Figure 3.13. I_D (V_{GS}) characteristics. Comparison
of the quantum model and the classic model

3.3.4.11.3. Conventional electric characteristics

Static electric characteristics of the DGMOS in question are therefore obtained
by resolving the "Poisson-Schrodinger" system in a self-consistent manner.

Figures 4.15–4.17 represent the potential energy and the electron density of the
DGMOS transistor for the bias voltages: $V_{DS} = 0.6$ V, $V_{GS} = 0.6$ V. The potential
energy barrier between the source and the channel is indicated here.

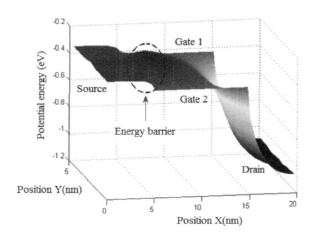

Figure 3.14. DGMOS potential energy

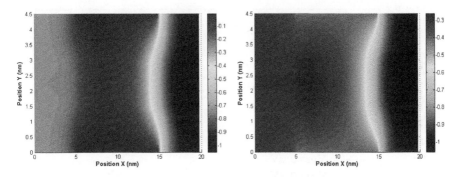

Figure 3.15. *DGMOS potential energy, V_{DS} = 0.8 V:*
(left) Vgs = 0.2 V, (right) V_{GS} = 0.6 V

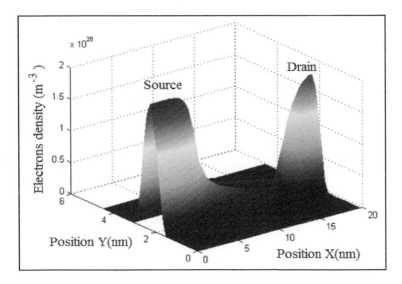

Figure 3.16. *Electron density: DGMOS*

Evolution of the potential energy and of the electron density of the double-gate structure for different gate voltages V_G are, respectively, represented in 2D in Figures 4.18(a) and (b), respectively.

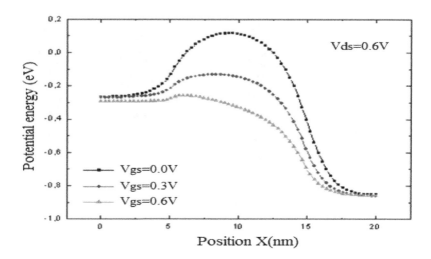

Figure 3.17. *Potential energy profiles along the channel for different gate polarizations*

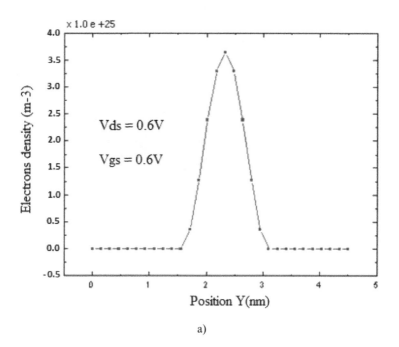

a)

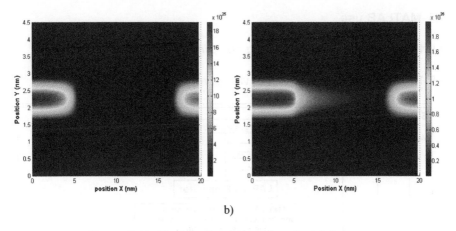

b)

Figure 3.18. *Electron density as a function of distance*
y in the middle of the channel for Vds = 0.8 V:
(left) Vgs = 0.2 V, (right) Vgs = 0.6 V

The decrease in the potential barrier is clearly seen with the increase in the gate voltage. At low gate voltage, the potential barrier of the channel does not allow for the passage of electrons coming from the source to the drain; in this case, the transistor is blocked. On the other hand, for higher gate voltage values, the potential barrier decreases and the electrons can pass from the source to the drain. The electron density along the channel for different gate voltage values confirms this process.

It can be seen that the density of electrons is strongly influenced by their quantification along the y axis. They are repelled by quantum effects, far from the Si/SiO$_2$. Indeed, the wave function tends toward zero at this interface.

3.4. Conclusion

This book first presents a presentation of the DGMOSFET and the numerical methods (Wigner/Monte-Carlo and Poisson/Schrödinger) to establish the electric characteristics of this device.

3.5. MATLAB use

3.5.1. *Computer-aided modelling and simulations: synopsis*

3.5.1.1. *From physics to MATLAB via appropriate equations*

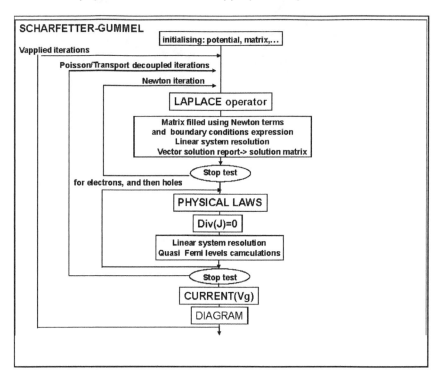

Figure 3.19. *The Scharfetter-Gummel algorithm*

POISSON'S EQUATION

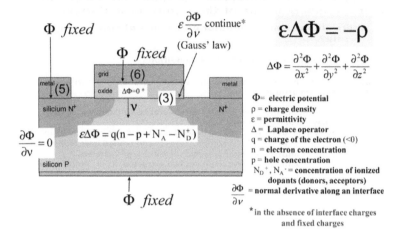

Figure 3.20. *Poisson equation applied to MOS*

DRIFT DIFFUSION MODEL

In each semiconductor

◆ Poisson equation for electrostatic potential

$$\varepsilon\Delta\Phi = q(n - p + N_A^- - N_D^+)$$

◆ current = diffusion current + drift current

◆ action of electric field $\vec{E} = -\vec{\nabla}\Phi$ $\vec{J_n} = -q\left(n\mu_n\vec{\nabla}\Phi - D_n\vec{\nabla}n\right)$

◆ D_n, D_p = diffusivity, μ_n, μ_p = mobility $\vec{J_p} = -q\left(p\mu_p\vec{\nabla}\Phi + D_p\vec{\nabla}p\right)$

◆ charge balance

◆ $\nabla\left(\vec{J_n} + \vec{J_p}\right) + q\dfrac{\partial}{\partial t}(p - n)) = 0$ ➡ $\left|\dfrac{\partial n}{\partial t} - \nabla\left[\dfrac{\vec{J_n}}{q}\right]\right. = G - R$

◆ steady state and without generation/recombination

$\nabla\vec{J_n} = \nabla\vec{J_p} = 0$ ⬅ $\left|\dfrac{\partial p}{\partial t} + \nabla\left[\dfrac{\vec{J_p}}{q}\right]\right. = G - R$

Figure 3.21. *"DDM" model*

... INTRODUCTION TO MORE PHYSICAL STUDIES...

Poisson and drift-diffusion equations are formulations that are independent of the nature of materials

... where do the semiconductor properties of silicon play a role?

Answer ... in boundary and interface conditions

Which constants?

What happens in a non-uniformly doped semiconductor's volume, in which all the contacts are grounded

The current is null, but the electric potential cannot be constant because an electric field stays around the junctions

⇒ 2 elecrochimical potentials appear, called pseudo levels of Fermi (n and p), which will be constant when no current is present , and related later to the statistics of occupation of the energy levels of the charges carriers

⇒ Value of polaryzarion applied extracted from solutions with constant Fermi levels.

Figure 3.22. *More physics?*

INSULATOR/SEMICONDUCTOR/METAL

INSULATOR   SEMICONDUCTOR   METAL

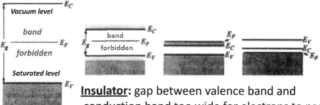

Insulator: gap between valence band and conduction band too wide for electrons to pass

Semiconductor: gap reduced, so electrons get through

In both cases, the E_F Fermi level is located between the energy levels of the 2 $E_C E_V$ bands

Metal: gap even smaller, or even non-existent ($E_V > E_C$), when it exists, the E_F Fermi level is beyond E_C

Figure 3.23. *MIS structure*

N-TYPE BAND DIAGRAM

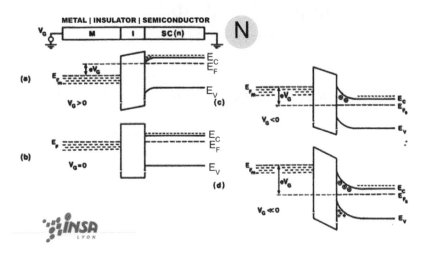

Figure 3.24. *MIS structure: N-type band diagram*

BAND DIAGRAM - P TYPE

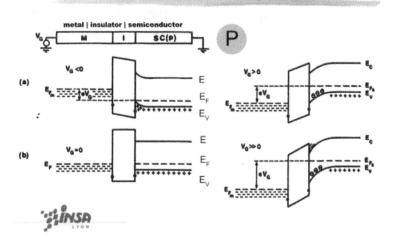

Figure 3.25. *MIS structure: P-type band diagram*

BOLTZMANN / POISSON STATISTICS

$$n = n_i\, e^{\frac{\Phi - \phi_n}{U_T}} \qquad p = n_i\, e^{\frac{\phi_p - \Phi}{U_T}}$$

$$\varepsilon \Delta\Phi = q\left(n_i e^{\frac{\Phi - \phi_n}{U_T}} - n_i e^{\frac{\phi_p - \Phi}{U_T}} + N_A^- - N_D^+ \right)$$

Φ = electric potential
ϕ_n, ϕ_n quasi Fermi levels

n_i = intrinsic concentration
$U_T = k_B T / q$
k_B = Boltzmann constant

Figure 3.26. *Boltzmann/Poisson statistics*

zero polarization solutions (thermodynamic equilibrium)

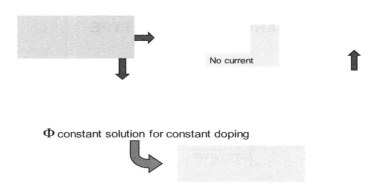

No current

Φ constant solution for constant doping

Figure 3.27. *Thermodynamic equilibrium*

BOUNDARY CONDITIONS: "POISSON"

⌘ Ohmic contacts $\qquad \Phi = V_{applied} \cdot U_T \text{Argsh}\left(\dfrac{N_D^+ - N_A^-}{2n_i}\right)$

⌘ Gate contacts $\qquad \Phi = V_{applied} + \Phi_{MS}$

⌘ Schottky contacts $\qquad \Phi = V_{applied} + \Phi_B$

⌘ Interfaces (Gauss' law) $\qquad \varepsilon_{SiO_2}\dfrac{\partial \Phi}{\partial \nu}\Big|_{SiO_2} - \varepsilon_{Si}\dfrac{\partial \Phi}{\partial \nu}\Big|_{Si} = q_{it}$

⌘ Axes of symetry $\qquad \dfrac{\partial \Phi}{\partial \nu} = 0$

Figure 3.28. *Boundary conditions*

BOUNDARY CONDITIONS: TRANSPORT

◆ Ohmic contacts:

- infinite carrier recombination
- electric neutrality

$$\begin{cases} n.p = n_i^2 \\ n - p + N_A^- - N_D^+ = 0 \end{cases}$$

◆ *Silicon/insulator interfaces and axes of symmetry*

$$-\overrightarrow{J_n}.\overrightarrow{i_n} = \overrightarrow{J_p}.\overrightarrow{i_p} = q.R_{surf}$$

Surface recombination velocity
(=0 on symmetry axes)

INSA
LYON

Figure 3.29. *Transport: boundary conditions*

MOS Capacitance

$Cox = \varepsilon_{SiO2} \cdot S / Tox$

$\varepsilon_{SiO2} = 35E\ 10^{-12}$

$S = 4E\ 10^{-8}$

$Tox = 2.6nm \Rightarrow Cox = 5{,}2.10^{-10}F$

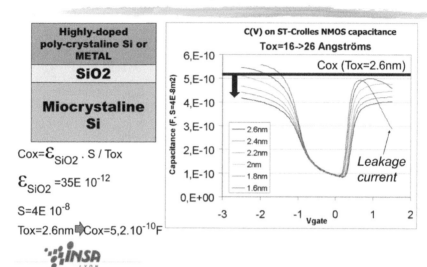

Figure 3.30. *MOS capacitance: C(V)*

DISCRETIZATION METHODS

- **Finite differences:**
  - the Laplacian operator is discretized (Taylor formula);
  - a regular mesh is needed.
- **Finite elements:**
  - *integration by parts*;
  - the functions are replaced by pieces of polynomials that connect.
- **In both cases:**
  - a mesh is needed;
  - the unknowns are the values of the solution at the nodes of the mesh;
  - *a very sparse matrix linear system is obtained.*

Figure 3.31. *Numerical methods*

FINITE DIFFERENCES METHOD

Principles

◆ Applying Taylor formula in each
spatial dimension

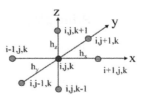

$\varepsilon(h_x)$, $\varepsilon(h_y)$ $\varepsilon(h_z)$ tends to 0 while h_x, h_y and h_z tend to 0.

$$\Delta_{ijk}(\Phi) = \frac{\Phi_{i+1jk} + \Phi_{i-1jk} - 2\Phi_{ijk}}{h_x^2} + \frac{\Phi_{ij+1k} + \Phi_{ij-1k} - 2\Phi_{ijk}}{h_y^2} + \frac{\Phi_{ijk+1} + \Phi_{ijk-1} - 2\Phi_{ijk}}{h_z^2}$$

◆ Example boundary condition: Φ_{ijk} fixed;

◆ Solution of a <u>nonlinear</u> system   ➡ Method of Newton-Raphson

➡ Linear system with sparse matrix

Figure 3.32. *Finite differences*

NEWTON-RAPHSON METHOD
FOR POISSON'S EQUATION

$$\varepsilon . f'' - a.e^f = 0$$ With the finite difference in 1D:

$$f''(x_i) \approx \frac{f(x_{i+1}) + f(x_{i-1}) - 2f(x_i)}{h^2}$$

➡ $\varepsilon \left(f_{i+1} + f_{i-1} - 2f_i \right) - a.h^2 e^{f_i} = F(f_{i-1}, f_i, f_{i+1}) = 0$

➡ $\frac{\partial F}{\partial f_{i-1}}\left(f_{i-1}^{k+1} - f_{i-1}^k\right) + \frac{\partial F}{\partial f_i}\left(f_i^{k+1} - f_i^k\right) + \frac{\partial F}{\partial f_{i+1}}\left(f_{i+1}^{k+1} - f_{i+1}^k\right) = -F\left(f_{i-1}^k, f_i^k, f_{i+1}^k\right)$ ⤸ k

$$\left(-2\varepsilon - a.h^2 e^{f_i}\right)\left(f_i^{k+1} - f_i^k\right)$$
$$+\varepsilon\left(f_{i+1}^{k+1} - f_{i+1}^k\right) + \varepsilon\left(f_{i-1}^{k+1} - f_{i-1}^k\right) = a.h^2.e^{f_i} + \varepsilon\left(2f_i^k - f_{i+1}^k - f_{i-1}^k\right)$$

➜ linear equation in the $\left(f_i^{k+1} - f_i^k\right)$
➜ for all nodes *i* of the mesh

Figure 3.33. *"Poisson": finite differences*

FINITE ELEMENT METHOD

Principles:

- ◆ Integration by parts (Green or Stokes formula);

- ◆ Integration domain meshing;

- ◆ Polynomial approximation on each element of:

- ◆ continuity at the boundary of elements,

- ◆ numerical integration;

- ◆ Very sparse matrix linear system.

Figure 3.34. *Finite elements*

FINITE ELEMENT METHOD FOR POISSON'S EQUATION

1D case

$$\varepsilon u''=F(u)$$ ➡️ $$\int_a^b (\varepsilon u''.w-F(u).w)dx=0$$

for all w in a suitably chosen functional space

Integration by parts

$$\int_a^b (-\varepsilon u'(x).w'(x)-F(u).w)dx+\varepsilon[u'(x).w(x)]_a^b=0$$

Values in a and b defined from boundary conditions

Finite elements ➡️

u and w replaced by piecewise and continuous polynomial functions, on a sampling of the interval $[a,b]$

$$w => w_i$$

$$u => \sum_{i=1}^{m} u_i w_i$$

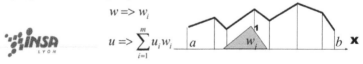

Figure 3.35. *"Poisson": finite elements*

VARIATIONAL FORMULATION OF POISSON'S EQUATION

Generalization for R^2 or R^3

$$\varepsilon\Delta u = F(u) \quad \Rightarrow \quad \int_{\Omega}\varepsilon\Delta u.w.d\Omega - \int_{\Omega}F(u).w.d\Omega = 0$$

Green formula:

$$\int_{\Omega}\varepsilon\vec{\nabla}u.\vec{\nabla}w.d\Omega + \int_{\Omega}\varepsilon\Delta u.w.d\Omega = \oint_{\partial\Omega}\varepsilon.\frac{\partial u}{\partial v}w.ds$$

$$-\int_{\Omega}\varepsilon\vec{\nabla}u.\vec{\nabla}w.d\Omega - \int_{\Omega}F(u).w.d\Omega = \oint_{\partial\Omega}\varepsilon.q_{it}.w.ds$$

for all *W* in a suitably chosen functional space

Figure 3.36. *"Poisson": variational method*

VARIATIONAL FORMULATION (continued)

❖ **Edge integrals** $\qquad \oint_{\partial\Omega}\varepsilon.\frac{\partial u}{\partial v}w.ds$

must include *all* interfaces between 2 materials

❖ **Transformation of edge integrals:**

⟹ zero flow conditions are expressed;
⟹ "test" functions w are chosen such that w = 0 on contacts;
⟹ Gauss' law is expressed on the silicon/insulator interfaces.

$$\oint_{\partial\Omega}\varepsilon.\frac{\partial u}{\partial v}w.ds = \oint_{\partial\Omega_{it}}\varepsilon.q_{it}.w.ds$$

Figure 3.37. *"Poisson" edge condition: variational method*

TRIANGULAR (2D) FINITE ELEMENTS

To solve:

$$-\int_{\Omega} \varepsilon \vec{\nabla} u.\vec{\nabla} w.d\Omega - \int_{\Omega} F(u).w.d\Omega = \oint_{\partial\Omega} \varepsilon.q_{it}.w.ds$$

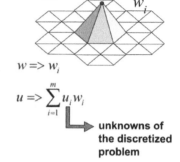

➡ more second derivatives;

➡ now, *u* and *w* can be replaced by simple basic functions, for example pisces of polynomials of degree 1 on each triangle, which connect;

➡ integrals are calculated numerically.

$$w => w_i$$

$$u => \sum_{i=1}^{m} u_i w_i$$

➡ unknowns of the discretized problem

Figure 3.38. *Finite elements: triangles*

DISCRETIZATION OF DIFFUSION DERIVATIVE EQUATIONS

- INOPERATIVE centered diagrams
  (finite differences and finite elements
  c.f. Poisson or pure diffusion equations)

➡ Risks of oscillations resulting from finite differences and finite elements:
  - ❖ Excessively coarse meshes;
  - ❖ Too high E.

➡ "Upstream" decentering (finite differences) or local conservation of current (finite volumes).

Figure 3.39. *"DDM": discretization*

FINITE VOLUME METHOD

Principle:

the local preservation of the stream is written on an offset mesh

in 1D **in 2D**

$$\int_{P_i} \nabla \overrightarrow{J_n} dP_i = \sum_j \oint_{S_{ij}} \overrightarrow{J_n}.\overrightarrow{i_n} ds$$

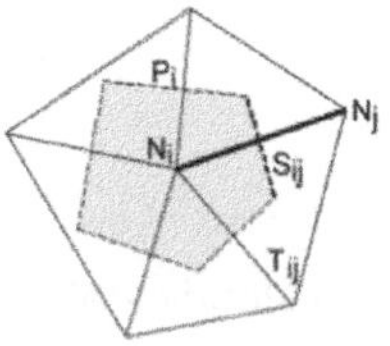

$$\int_{x_{i-1/2}}^{x_{i+1/2}} \frac{dJ_n}{dx} dx = \left[J_n \right]_{x_{i-1/2}}^{x_{i+1/2}} = J_n(x_{i+1/2}) - J_n(x_{i-1/2})$$

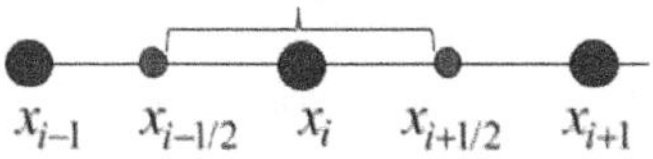

$$J_n(x) = -q.n.\mu_n \frac{d\phi_n}{dx}$$

Trick!

$$J_n(x).e^{-\Phi/UT} = -q.n.\mu_n \frac{d\phi_n}{dx}.e^{-\Phi/UT} = -q.n_i.\mu_n \frac{d\phi_n}{dx}.e^{-\phi_n/UT} = q.n_i.\mu_n.UT.\frac{d(e^{-\phi_n/UT})}{dx}$$

INSA LYON

... we integrate on $\left[-x_i, x_{i+1} \right]$ **at constant** J_n

Figure 3.40. *Finite volumes: triangles*

1D discretization of transport equations

$$J_n(x).e^{-\Phi/UT} = -q.n.\mu_n \frac{d\phi_n}{dx}.e^{-\Phi/UT} = -q.n_i.\mu_n \frac{d\phi_n}{dx}.e^{-\phi_n/UT} = q.n_i.\mu_n.UT.\frac{d(e^{-\phi_n/UT})}{dx}$$

We integrate on $\left[x_i, x_{i+1} \right]$ assuming J_n and μ_n are constants

Exact integration of the term $e^{-\Phi/UT}$ for linear Φ

$$\Rightarrow J_n(x_{i+1/2}) \int_{x_i}^{x_{i+1}} e^{-\Phi/UT} dx = J_n(x_{i+1/2}).\Delta x.UT \left[\frac{e^{-\Phi_{i+1}/UT} - e^{-\Phi_i/UT}}{\Phi_{i+1} - \Phi_i} \right]$$

$$= q.n_i.\mu_n.UT \left[e^{-\phi_n/UT} \right]_{x_i}^{x_{i+1}} = q.\mu_n.UT \left[n.e^{-\Phi/UT} \right]_{x_i}^{x_{i+1}}$$

Ditto on $\left[x_{i-1}, x_i \right]$

We can bring all calculations back to the nodes x_{i-1}, x_i, x_{i+1} ....

Figure 3.41. *Transport in 1D*

Discretization of transport equations

$$J_n(x_{i+1/2})=\frac{q.\mu_n.UT\left(\Phi_{i+1}-\Phi_i\right)\left(n(x_{i+1}).e^{-\Phi_{i+1}/UT}-n(x_i).e^{-\Phi_i/UT}\right)}{\Delta x\left(e^{-\Phi_{i+1}/UT}-e^{-\Phi_i/UT}\right)}$$

In steady state, without generation–recombination

$$Div(J_n)=0 \implies J_n(x_{i-1/2})=J_n(x_{i+1/2})$$

Expression using the Bernouilli function $B(u)=\dfrac{u}{e^u-1}$

$B_{ii+1}=B((\Phi(x_{i+1})-\Phi(x_i))/UT)$ with $n_i=n(x_i)$, $n_j=n(x_{i+1})$

$$\implies \frac{\mu_{n,i+1/2}}{x_{i+1}-x_i}\left(n_{i+1}.B_{ii+1}-n_i.B_{i+1i}\right)=\frac{\mu_{n,i-1/2}}{x_i-x_{i-1}}\left(n_i.B_{i-1i}-n_{i-1}.B_{ii-1}\right)$$

Linear relationship on carrier concentrations at nodes

Figure 3.42. *Transport: 1D discretization*

2D FINITE VOLUME METHOD

$$\int_{P_i}\nabla\vec{J}_n dP_i=\sum_j\oint_{S_{ij}}\vec{J}_n.\vec{i}_n ds$$

$$-\sum_j\frac{\|S_{ij}\|}{\|N_iN_j\|}D_n\left(n_iB_{ji}-n_jB_{ij}\right)=\int_{P_i}(\frac{\partial n_i}{\partial t}-G+R)dP_i$$

$$-\sum_j\frac{\|S_{ij}\|}{\|N_iN_j\|}D_p\left(p_iB_{ij}-p_jB_{ji}\right)=\int_{P_i}(\frac{\partial p_i}{\partial t}-G+R)dP_i$$

$B(u)=\dfrac{u}{e^u-1}$ $B_{ij}=B(u(N_j)-u(N_i))$, $n_i=n(N_i)$, $n_j=n(N_j)$

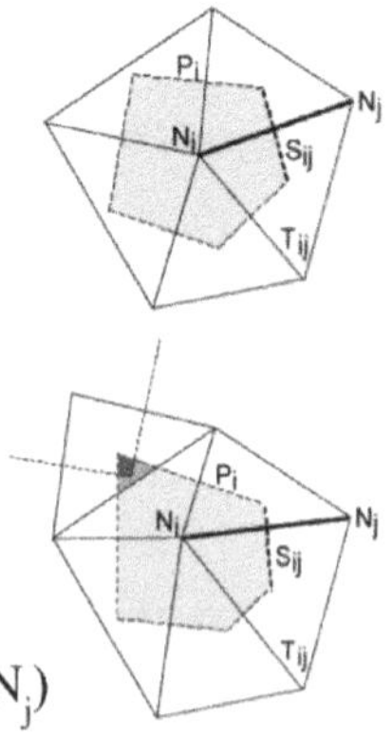

Figure 3.43. *Finite volumes: 2D*

TIME DISCRETIZATION

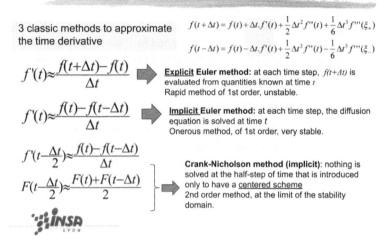

3 classic methods to approximate the time derivative

$$f(t+\Delta t) = f(t) + \Delta t . f'(t) + \frac{1}{2}\Delta t^2 f''(t) + \frac{1}{6}\Delta t^3 f'''(\xi_+)$$

$$f(t-\Delta t) = f(t) - \Delta t . f'(t) + \frac{1}{2}\Delta t^2 f''(t) - \frac{1}{6}\Delta t^3 f'''(\xi_-)$$

$$f'(t) \approx \frac{f(t+\Delta t) - f(t)}{\Delta t}$$

Explicit Euler method: at each time step, $f(t+\Delta t)$ is evaluated from quantities known at time t
Rapid method of 1st order, unstable.

$$f'(t) \approx \frac{f(t) - f(t-\Delta t)}{\Delta t}$$

Implicit Euler method: at each time step, the diffusion equation is solved at time t
Onerous method, of 1st order, very stable.

$$f'(t-\frac{\Delta t}{2}) \approx \frac{f(t) - f(t-\Delta t)}{\Delta t}$$

$$F(t-\frac{\Delta t}{2}) \approx \frac{F(t) + F(t-\Delta t)}{2}$$

Crank-Nicholson method (implicit): nothing is solved at the half-step of time that is introduced only to have a <u>centered scheme</u>
2nd order method, at the limit of the stability domain.

INSA
LYON

Figure 3.44. *Temporal dictation*

SMALL SIGNAL ANALYSIS

Example: Diffusion equation (linear)

$$\frac{\partial \Phi}{\partial t} = D.\Delta \Phi$$
$$\Phi(x,t) \in R$$

Sinusoidal signal

$$\Phi = \widetilde{\Phi} e^{i\omega t} + \overline{\widetilde{\Phi}} e^{-i\omega t} = 2|\widetilde{\Phi}|\cos(\omega t + \varphi)$$

$$\frac{\partial \Phi}{\partial t} = i\omega(\widetilde{\Phi} e^{i\omega t} - \overline{\widetilde{\Phi}} e^{-i\omega t})$$

$$D.\Delta \Phi = D.\Delta\widetilde{\Phi} e^{i\omega t} + D.\Delta\overline{\widetilde{\Phi}} e^{-i\omega t}$$

$$i\omega\widetilde{\Phi} = D.\Delta\widetilde{\Phi}$$
$$\widetilde{\Phi}(x) \in C$$

For any t

INSA
LYON

Figure 3.45. *Diffusion equation: small signal analysis*

3.5.1.2. *Modeling of a PN Junction*

Here, we propose a simulation method, which is obtained thanks, in particular, to the solution of Poisson equations using the Newton algorithm, as well as the use of the Scharfetter and Gummel algorithm for electron and hole transport.

This section briefly explains the operation of the program that models a PN junction and also gives an overview of its theoretical foundations.

Modeling the PN junction (diode) was solved by implementing the Scharfetter–Gummel algorithm. This algorithm was designed in 1969 by Hermann Gummel and Donald Lee Scharfetter, and allows for numerical resolution of equations that govern the diode potential and charge.

PN junction: operation (recap)

A PN junction or diode is an electronic device consisting of the juxtaposition of two regions of the same semiconductor single crystal, which have been doped differently. The operating principle of the junction can be easily explained from the microscopic point of view.

Everyone knows that when two atoms are distant, their electronic states are represented by atomic orbitals. When atoms get closer, atomic states couple to give rise to molecular orbits: one is a low-energy orbit, called a binder, and another is a high-energy orbit, called an anti-binder. A covalent bond is then referred to; this is distinctive from the elements belonging to column IV of the period table, namely C, Si, Ge and Sn.

The two sets of orbits constitute, energetically, permitted bands for electrons. For a given interatomic distance, the population of permitted bands is a function of the splits of the s-p orbits in the isolated atom, and binder–anti-binder in chemical bonds of the crystal. The atom density is about 10^{22} cm^{-3}.

The conduction phenomenon is the result of the displacement of electrons within bands of binding orbits. Thus, an insulator has only totally solid or totally empty bands. Therefore, a semiconductor has bands which allow a limited movement of electrons (i.e. they have a filling ratio of less than 25% or more than 75%).

Allowed bands are classified into valence bands (those that are full) and conduction bands (those that are empty). There is a forbidden band between them, called a gap, with which an energy Eg of electron thermal agitation is associated. Eg is the energy that must be supplied to an electron in order for it to change from a valence band to a conduction band: Eg = 1.12 eV for silicon.

Semiconductor (Si) single crystals are doped to increase or decrease the electron density in the permitted bands. Doping is the ion implantation (I^2) of different atoms in the crystal lattice. This can be:

– N type, if implanted atoms play the role of donors, as in the case of P or As. An electron is released by thermal agitation (energy required of the order of meV) and occupies an orbit of the conduction band;

– P type, if implanted atoms act as acceptors, as in the case of B, Al, or Ga. An electron passes by thermal agitation from an orbit of the valence band to the implanted atom. A hole is then created.

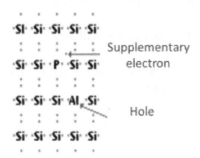

Figure 3.46. *P-doped Si matrix*

Electronic neutrality in the crystal remains valid, and *id* is:

$$n + N_a^- = p + N_d^+ \tag{3.99}$$

where n, p, N_a^- and N_d^+ correspond to the density of electrons, holes, acceptors and donors, respectively, resulting from thermal excitation. Henceforth, we will call free carriers electrons and holes, and ionized carriers acceptors and donors.

At the equilibrium of the semiconductor, the latter is characterized by the intrinsic density of carriers n_i. Thus, we have:

$$n = p = n_i \tag{3.100}$$

– Discretization of equations

We want to numerically model equations that define potential (φ) and pseudo-potentials (φ_n and φ_p) in the semiconductor.

The system of equations that we will use is as follows:

$$div(M_n \cdot \exp(\phi) \cdot \overrightarrow{grad}(\psi_n)) = GR \tag{3.101}$$

$$div(M_p \cdot \exp(-\phi) \cdot \overrightarrow{grad}(\psi_p)) = -GR \tag{3.102}$$

$$div(\overrightarrow{grad}(\phi)) = \exp(\phi) \cdot \psi_n - \exp(-\phi) \cdot \psi_p - DOP \tag{3.103}$$

By solving the system, we find the expressions of $\psi_n(x, y)$ as a function of $\exp(\phi)$.

The same calculation is made for $\psi_p(x, y)$ from:

$$div(M_p \cdot \exp(-\phi) \cdot \overrightarrow{grad}(\psi_p)) = -GR \qquad [3.104]$$

Thus, for each point (x, y), values found for ψ_n and ψ_p are entered into the right-hand side of equation [3.104]:

$$div(\overrightarrow{grad}(\phi)) = \exp(\phi) \cdot \psi_n - \exp(-\phi) \cdot \psi_p - DOP \qquad [3.105]$$

This way, we have $div(\overrightarrow{grad}(\phi))$ as a function of the partial derivatives of φ, and by a new application of the finite difference method we find the value of φ at points (x, y) using:

– MATLAB programming;

– resolution methodology via finite elements;

– triangular finite elements ρ^1.

Here, we study a finite-element approximation ρ^1 of a generic problem over any Ω two-dimensional domain.

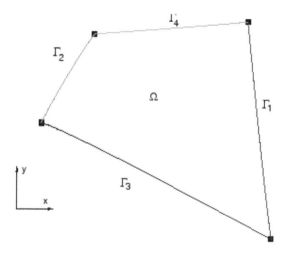

Figure 3.47. *An example of a computing domain – M. Buffatt*

Boundary $G = d\,W$ is made up of four disjoint sub-boundaries:

$$\Gamma = \Gamma_1 \cup \Gamma_2 \cup \Gamma_3 \cup \Gamma_4$$

Each boundary complies with one type of boundary condition applied to solution u(x,y):

1) G_1: homogeneous Dirichlet boundary: namely: $u_{G1} = 0$;

2) G_2: non-homogeneous Dirichlet boundary: that is, $u_{G2} = u_e$;

3) G_3: homogeneous Neumann boundary: that is, $\left(\dfrac{\partial u}{\partial n}\right)_{r_3} = 0$;

4) G_4: Fourier boundary: that is, $-K\left(\dfrac{\partial u}{\partial n}\right)_{r_4} = \beta u_{r_4} + \Phi_0$.

The problem to be studied can be written as:

$$\begin{cases} -\dfrac{\partial}{\partial x}\left(K\dfrac{\partial u}{\partial x}\right) - \dfrac{\partial}{\partial y}\left(K\dfrac{\partial u}{\partial y}\right) + \alpha u\left(x,y\right) = f\left(x,y\right) \text{ on } \Omega \\[2mm] u\Gamma_1 = 0, u\Gamma_2 = u_e, \left(\dfrac{\partial u}{\partial n}\right)\Gamma_3 = 0, -K\left(\dfrac{\partial u}{\partial n}\right)\Gamma_4 = \beta u\Gamma_4 + \phi_0 \end{cases} \qquad [3.106]$$

The equation considered here is, for example, a convection/conduction equation with a source term that is a function of the solution. K is the diffusion coefficient (K > 0), a is the coefficient of the linear source term (a > 0) in u(x,y) and f(x,y) is the source term independent of u(x,y).

– Weak formulation

The weak formulation of [3.106] is obtained by multiplying by a test function u(x,y) and then integrating over the entire Ω domain. The second derivative term is then *partially* integrated via a Green formulation. Considering boundary conditions and interpreting the test function v(x,y) as a variation of the solution u(x,y), the weak formulation is written as follows:

$$\begin{cases} \textbf{Find } \; u(x,y) \text{ t.q. } u_{\Gamma_1} = 0, u_{\Gamma_2} = u_e \\[1mm] \int_\Omega K\left(\dfrac{\partial u}{\partial x}\dfrac{\partial v}{\partial x} + \dfrac{\partial u}{\partial y}\dfrac{\partial v}{\partial y}\right) d\omega + \int_\Omega \alpha u.v\, d\omega + \int_{\Gamma_4}\beta u.v\, d\gamma = \int_\Omega f v\, d\omega - \int_{\Gamma_4}\phi_0 v\, d\gamma \\[1mm] \forall v(x,y) \text{ t.q. } v_{\Gamma_1} = 0, v_{\Gamma_2} = 0 \end{cases} \qquad [3.107]$$

To obtain the approximate solution u_k, we choose a **W** mesh, as well as an approximation of the solution on this mesh.

– Finite element mesh: interpolation ρ^1

The computing domain Ω is (often) divided into triangles (see Figure 3.48). There are many examples of software; here, in this case, MATLAB, Python, and Scilab.

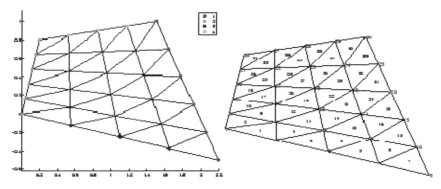

Figure 3.48. *Finite element mesh ρ^1*

Here is some information about this mesh:

1) number of nodes nn, number of elements ne;

2) coordinates (x_i, y_i) of each node M_i of the mesh;

3) the numbers of the vertices of each element e_k (see connection table): $tbc_{k,1}, tbc_{k,2}, tbc_{k,3}$;

4) for each node M_i, there is information *frt* that specifies whether the node is internal (*frt$_i$* = 0), or whether we are on a boundary G_1 (*frt$_i$* = 0);

5) for each element e_k, information on the region, reg$_k$, specifying the domain number to which the element belongs. By default, there is only one domain and then reg$_k$ = 1, regardless of the element.

This information is saved as a table in a "geometry" file, with an .msh (mesh) extension. This file is written in ASCII and is structured by line as follows:

Line	Field 1	Field 2	Field 3	Field 4

Figure 3.49. *Mesh file*

The mesh file in Figure 3.49 contains, for example, the following values:

30 40 mesh_triangle_P1

0.000000 0.000000 2

0.550000 -0.125000 3

1.100000 -0.250000 3

....

0.850000 0.900000 4

1.175000 0.950000 4

1.500000 1.000000 1

1 2 7 1

6 1 7 1

....

24 25 30 1

29 24 30 1

With MATLAB, we use a data structure for structure-type geometry, which will contain the following fields:

– [G.nn] number of nodes of geometry G: integer;

– [G.ne] number of elements of geometry G: integer;

– [G.dim] dimension of space (=2 in 2D): integer;

– [G.ddl] number of degrees of freedom per element (=3 for ρ^1 2D elements): integer;

– [G.X] table of the coordinates (x,y) of nodes: real G.nn*G.dim table;

– [G.Tbc] connection table: G.ne*G.ddl integers table;

– [G.Frt] boundary number by nodes: integer G.nn table;

– [G.Reg] number of region by elements: G.ne integers table.

The MATLAB Lecture.m function reads a "finite element" geometry, initializing the geometry structure variable G.

– MATLAB program: Reading geometry: Lecture.m

```
function [G]=Read(file)
% mesh reading
% opening file
fid=fopen(fichier,'r');
% dimension reading
[L,count]=fscanf(fid,'%d %d %s',3);
G.nn=L(1); G.ne=L(2);
% geometry structure initialization
G.dim=2; G.ddl=3;
G.X=zeros(G.nn,G.dim); G.Frt=zeros(G.nn,1);
G.Tbc=zeros(G.ne,G.ddl); G.Reg=zeros(G.ne,1);
% coordinates reading
for i=1:G.nn
   [L,count]=fscanf(fid,'%f %f %d',3);
   G.X(i,1:2)=[L(1) L(2)];
   G.Frt(i)=L(3);
end;
% disconnection table
for l=1:G.ne
   [L,count]=fscanf(fid,'%d %d %d %d',4);
   G.Tbc(l,1:3)=[L(1) L(2) L(3)];
   G.Reg(l)=L(4);
end;
fclose(fid);
return;
```

Then, still via MATLAB, to read the stored geometry in the mesh.msh file, we write:

➔ G1=Lecture ('maillage.msh')

G1 =

   nn: 30
   ne: 40
   dim: 2
   ddl: 3
    X: [30x2 double]
   Frt: [30x1 double]
   Tbc: [40x3 double]
   Reg: [40x1 double]

3.5.1.2.1. Finite element approximation

An approximation $u_k(x,y)$ on a finite element mesh is a linear combination of the nodal values $\{u_i\}_{i=1,\,nn}$ and the nodes of the mesh. Coefficients of the latter are basic functions $\Phi_i(x,y)$.

$$u^h(x,y) = \sum_{i=1}^{nn} u_i \Phi_i(x,y) \quad \text{with} \quad u_i = u^h(x_i, y_i) \tag{3.108}$$

The $\{F_i\}$ constitutes a local base, namely, a function $F_i(x,y)$ is non-zero only on a reduced part of the mesh: the support of node M_i, that is, all elements e_l having node M_i as vertex (M_i). In addition, they verify the following properties:

1) the basic function $F_i(x,y)$ is equal to 1 at the node i and 0 on all the other nodes:

$$\Phi_i(x_j, y_j) = \delta_{ij} \tag{3.109}$$

The basic function $F_i(x,y)$ is not zero only on its support

$$Sup_i = \{\, Ue_k / Mi \in e_k \,\} \tag{3.110}$$

$$N_i(x,y) \neq 0 \quad \text{if} \quad (x,y) \subset Sup_i \tag{3.111}$$

2) $F_i(x,y)$ is orthogonal to almost all the other functions F_j, in particular those which are associated with nodes j not next to i (e.g. $Sup_i \cap Sup_j = 0$):

$$\Phi_i(x,y) * \Phi_j(x,y) \begin{cases} = 0 & \text{if } Sup_i \cap Sup_j = 0 \\ = 0 & \text{if } (x,y) \notin (Sup_i \cap Sup_j) \\ \neq 0 & \text{if } (x,y) \in (Sup_i \cap Sup_j) \end{cases} \tag{3.112}$$

A typical example of basic functions (associated with nodes 7, 12, 24 of the mesh of Figure 3.49) is plotted in Figure 3.51, with their supports; the basic functions F_7 and F_{24} are orthogonal (the $F_7 * F_{24}$ product is always zero), as well as F_{12} and F_{24}, and the $F_7 * F_{12}$ product is non-zero only in the two elements (e_9, e_{12}).

To calculate a basic function on an element e_k of the mesh, a transformation of this element toward a reference element ê is used. On the latter, the basic function associated with one of the vertices $\{ S_q \}_{q=1,3}$ of the element e_k thus merged with one of the form functions $\{N_q(z,h\}$, which are associated with vertices $\{\hat{S}_q\}$ of the element of ê.

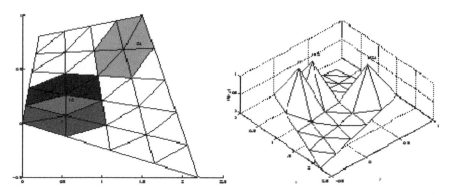

Figure 3.50. *Basic functions F_7, F_{12}, and F_{24} with their supports*

– Discrete weak formulation

The approximate solution u^h of equation [3.108] can be written as:

$$u^h(x,y) = \sum_{i-1}^{nn} u_i \Phi_i(x,y) \qquad\qquad [3.113]$$

In addition, it must verify the conditions at the Dirichlet boundaries: $u^h_{\Gamma_1} = 0$, $u^h_{\Gamma_2} = u_e$.

The conditions set the nodal value of u^h on all nodes belonging to the boundary Γ_1 ou Γ_2, that is:

$$u_l = 0 \text{ if } M_l \in \Gamma_1, u_l = u_e \text{ if } M_l \in \Gamma_2 \qquad\qquad [3.114]$$

Let nf_1 be the number of nodes on boundary Γ_1, and let nf_2 be the number of nodes on boundary Γ_2; the number of degrees of freedom nl of u^h is equal to nl = nn – nf_1 – nf_2. By denoting $\overline{\Omega}_d = W.- G_{1.} - G_{2.}$, the solution can be written in the following form:

$$u^h(x,y) = \sum_{j \in \Omega d} u_j \Phi_j(x,y) + \sum_{l \in \Gamma_2} u_e \Phi_l(x,y) \qquad\qquad [3.115]$$

The associated test functions $v^h(x,y)$ are variations of u_h, which must therefore cancel out Dirichlet boundaries G_1 and G_2. They are written in the following form:

$$u^h(x,y) = \sum_{i=1}^{nn} u_i \, \Phi_i(x,y)$$

[3.116]

They are a linear combination of the n_l basic functions F_i ($i \ \overline{\Omega}_d$)= W. By replacing the exact solution u by the expression [3.116] of u^h and the test function v by one of the preceding basic functions F_i; hence, the discrete weak formulation:

$$\sum_{j \in \overline{\Omega}d} u_j \left(\underbrace{\int_\Omega K \left(\frac{\partial \Phi_j}{\partial x} \frac{\partial \Phi_i}{\partial x} + \frac{\partial \Phi_j}{\partial y} \frac{\partial \Phi_i}{\partial y} \right) d\omega + \int_\Omega \alpha \Phi_j \cdot \Phi_i d\omega}_{\text{Generic term of A}} + \underbrace{\int_{\Gamma_4} \beta \Phi_j \cdot \Phi_i d\gamma}_{\text{CL term on A}} \right)$$

$$\underbrace{\int_\Omega f \Phi_i d\omega}_{\text{Generic term of B}} - \underbrace{\int_{\Gamma_4} \phi_0 \Phi_i d\gamma}_{\text{CL term on B}}$$ [3.117]

$$\sum_{l \in \Gamma_2} u_l \underbrace{\left(\int_\Omega K \left(\frac{\partial \Phi_l}{\partial x} \frac{\partial \Phi_i}{\partial x} + \frac{\partial \Phi_l}{\partial y} \frac{\partial \Phi_i}{\partial y} \right) d\omega + \int_\Omega \alpha \Phi_l \cdot \Phi_i d\omega + \int_{\Gamma_4} \beta \Phi_l \cdot \Phi_i d\gamma \right)}_{\text{CL term on B}}$$

– The relationship [3.117] written for the nl basic functions N_i belonging to $\overline{\Omega}_d$ is a linear system of unknown nl u_j. We build, ab initio, a linear system e for all these nodal values $\{u_q\}_{i=1,nn}$; then, we apply the conditions to boundaries.

The coefficients of matrices A and the second term B are written as follows:

$$\mathbf{A}_{ij} = \int_\Omega K \left(\frac{\partial \Phi_j}{\partial x} \frac{\partial \Phi_i}{\partial x} + \frac{\partial \Phi_j}{\partial y} \frac{\partial \Phi_i}{\partial y} \right) d\omega + \int_\Omega \alpha \Phi_j \cdot \Phi_i d\omega$$

$$\mathbf{B}_i = \int_\Omega f \Phi_i d\omega$$

[3.118]

To calculate the latter, we calculate element by element via determining matrices and the second elementary parts on each element e_k:

$$A_{ij} = \sum_{k=1}^{ne} \left(\underbrace{\int_{e_k} K \left(\frac{\partial \Phi_j}{\partial x} \frac{\partial \Phi_i}{\partial x} + \frac{\partial \Phi_j}{\partial y} \frac{\partial \Phi_i}{\partial y} \right) d\omega + \int_{e_k} \alpha \Phi_j \cdot \Phi_i d\omega}_{\text{Elementary matrix}} \right) \quad [3.119]$$

$$B_i = \sum_{k=1}^{ne} \left(\underbrace{\int_{ek} f \Phi_i d\omega}_{\text{Second elementary member}} \right)$$

– Elementary matrices calculation ρ^1

As a function of the properties of the basic functions, on an element e_k, it is necessary to calculate 3×3 elementary integrals involving the three basic functions associated with the three vertices of the element. By denoting by $\{n_1, n_2, n_3\}$ the numbers of these three vertices, it is then necessary to calculate the following elementary matrix 3×3 on an element e_k:

$$A_{pq}^k = \int_{ek} K \left(\frac{\partial \Phi_{n_q}}{\partial x} \frac{\partial \Phi_{n_p}}{\partial x} + \frac{\partial \Phi_{n_q}}{\partial y} \frac{\partial \Phi_{n_p}}{\partial y} \right) d\omega + \int_{ek} \alpha \Phi_{n_q} \cdot \Phi_{n_p} d\omega \, (p,q = 1,3) \quad [3.120]$$

Knowing that coefficients K and a are constant on element e_k, this matrix is the sum of an elementary stiffness matrix $\mathbf{K}^k$ and a mass matrix $\mathbf{M}^k$:

$$A_{pq}^k = K\mathbf{K}_{pq}^k + \alpha\mathbf{M}_{pq}^k \quad [3.121]$$

These two matrices are calculated via the change in variables to the reference element e_k.

3.5.1.2.2. Elementary stiffness matrix

Here is the expression for the stiffness matrix calculated on the reference element:

$$\mathbf{K}_{pq}^k = \int_0^1 \int_0^{1-\xi} \left[\left(\left(\mathbf{J}_k^{-1} \right)^t \begin{bmatrix} \dfrac{\partial N_p}{\partial \xi} \\ \dfrac{\partial N_p}{\partial \eta} \end{bmatrix} \right)^t \cdot \left(\left(\mathbf{J}_k^{-1} \right)^t \begin{bmatrix} \dfrac{\partial N_q}{\partial \xi} \\ \dfrac{\partial N_q}{\partial \eta} \end{bmatrix} \right) \right] \det \left(\mathbf{J}_k \right) d\eta \, d\xi \quad [3.122]$$

This matrix depends on three coefficients, which can be written in vector form as:

$$\mathbf{K}_{22}^k = \frac{\left\|\overrightarrow{S_1S_3}\right\|^2}{4area_k}, \mathbf{K}_{33}^k = \frac{\left\|\overrightarrow{S_2S_1}\right\|^2}{4area_k}, \mathbf{K}_{23}^k = \frac{\overrightarrow{S_1S_3}\cdot\overrightarrow{S_2S_1}}{4area_k}, area_k = \frac{1}{2}\left\|\overrightarrow{S_1S_3}\otimes\overrightarrow{S_2S_1}\right\| \quad [3.123]$$

which is referred to in the K_K expression:

$$\mathbf{K}^k = \begin{bmatrix} K_{22}^k + K_{33}^k + 2K_{23}^k & -K_{23}^k - K_{22}^k & -K_{23}^k - K_{33}^k \\ -K_{23}^k - K_{22}^k & K_{22}^k & K_{23}^k \\ -K_{23}^k - K_{33}^k & K_{23}^k & K_{33}^k \end{bmatrix} \quad [3.124]$$

The MatrixRigidite function (program below) directly implements previous relationships in MATLAB. In addition, the fact that the vector product $\overrightarrow{S_1S_3}\otimes\overrightarrow{S_2S_{31}}$ has a single non-zero component, which is along the z axis, has been used; it is positive if the vertices are given in trigonometric order in the connection table (which is the case).

– MATLAB program: Elementary stiffness matrix: MatrixRigidite.m

```
function [Ke]=MatriceRigidite(G,k)
% calculation of the stiffness matrix of element k
n=G.Tbc(k,:); % number of vertices of element k
S21=G.X(n(1),:)-G.X(n(2),:);
S13=G.X(n(3),:)-G.X(n(1),:);
Area=0.5*(S13(1)*S21(2)-S13(2)*S21(1));
K22=S13*S13'/(4*Area);
K33=S21*S21'/(4*Area);
K23=(S13*S21')/(4*Area);
Ke = [K22+K33+2*K23,-K23-K22,-K23-K33;
   -K23-K22, K22, K23;
   -K23-K33, K23, K33];
return;
```

– Elementary mass matrix

Similarly, the change in variable makes it possible to obtain the mass matrix:

$$\mathbf{M}_{pq}^k = \int_0^1 \int_0^{1-\xi} N_q(\xi,\eta) N_p(\xi,\eta) \det(\mathbf{J}_k)\, d\eta\, d\xi \quad [3.125]$$

On the reference element, the calculation of this matrix is simple, because the determinant of the Jacobian is constant, $\det(J^k) = 2.aire.e_k$, and the functions of forms L_q and L_p are simple polynomials in () (z, h). Considering the expression of these polynomials and symmetry properties, it is only necessary to calculate two integrals:

$$\int_0^1 \int_0^{1-\xi} \eta\xi \, d\eta \, d\xi = \frac{1}{24}, \int_0^1 \int_0^{1-\xi} \xi^2 \, d\eta \, d\xi = \frac{1}{12}$$ [3.126]

Indeed, the symmetry properties lead to the following:

$$\int_0^1 \int_0^{1-\xi} (N_1)^2 \, d\eta \, d\xi = \int_0^1 \int_0^{1-\xi} (N_2)^2 \, d\eta \, d\xi = \int_0^1 \int_0^{1-\xi} (N_3)^2 \, d\eta \, d\xi = \int_0^1 \int_0^{1-\xi} \xi^2 \, d\eta \, d\xi$$ [3.127]

$$\int_0^1 \int_0^{1-\xi} N_1 N_2 \, d\eta \, d\xi = \int_0^1 \int_0^{1-\xi} N_1 N_3 \, d\eta \, d\xi = \int_0^1 \int_0^{1-\xi} N_2 N_3 \, d\eta \, d\xi = \int_0^1 \int_0^{1-\xi} \eta\xi \, d\eta \, d\xi$$ [3.128]

Hence, the elementary mass matrix:

$$\mathbf{M}^k = \frac{area_k}{12} \begin{bmatrix} 2 & 1 & 1 \\ 1 & 2 & 1 \\ 1 & 1 & 2 \end{bmatrix}$$ [3.129]

whose MATLAB MassMatrix function is written below.

– MATLAB program: Elementary mass matrix: MassMatrix.m

```
function [Me]=MassMatrix(G,k)
% calculation of the stiffness matrix of element k
n=G.Tbc(k,:); % number of vertices of element k
S21=G.X(n(1),:)-G.X(n(2),:);
S13=G.X(n(3),:)-G.X(n(1),:);
Area=0.5*(S13(1)*S21(2)-S13(2)*S21(1));
Me=Area/12*[2,1,1; 1,2,1; 1,1,2];
return;
```

3.5.2. *Calculation of the second elementary member* ρ^1

For each element e_k, three elementary integrals associated with each of the three vertices of the element must be calculated. With the same notations as before, these integrals are written as follows:

$$\mathbf{B}_p^k = \int_{ek} f \cdot \Phi_{n_p} d\omega (p = 1,3) \tag{3.130}$$

To calculate these integrals, function f' (x,y) is replaced by its interpolation $f^h(x,y)$ on the finite element mesh:

$$f^h(x,y) = \sum_{i=1}^{nn} f_i \Phi_i(x,y) \tag{3.131}$$

This makes it possible to write the integrals in the following form:

$$\mathbf{B}_p^k = \sum_{q=1}^{3} f_{n_q} \left(\int_{e_k} \Phi_{n_q} \cdot \Phi_{n_p} d\omega \right) (p = 1,3) \tag{3.132}$$

On the element e_k, the interpolation of f, f^h is written as: $f^h(x,y) = \sum_{q=1}^{3} fn_q \cdot Nn_q(x,y)$.

This is the product of vector $F_k = [fn_1, fn_2, fn_3]$ of nodal values by the elementary mass matrix M^k. The second elementary part is then written as:

$$\mathbf{B}^k = \mathbf{M}^k \mathbf{F}^k = \frac{area_k}{12} \begin{bmatrix} 2 & 1 & 1 \\ 1 & 2 & 1 \\ 1 & 1 & 2 \end{bmatrix} \begin{bmatrix} f_{n1} \\ f_{n2} \\ f_{n3} \end{bmatrix} \tag{3.133}$$

and the MATLAB SmbElement function is given below.

– MATLAB Program: Second Elementary Part: SmbElement.m

```
function [Be]=SmbElement(G,F,k)
% calculation of the second elementary part (F nodal values of f)
n=G.Tbc(k,:); % number of vertices of element k
Fk=F(n);
Mk=MassMatrix(G,k);
Be=Mk*Fk;
return;
```

– Assembly

The assembly of matrix **A** and the second term **B** consists of calculating elementary mass and stiffness matrices, and the second elementary part, then putting its values at the "right" places in the matrix and the second overall part.

The assembly algorithm (2) indicates the principle of assembly.

Algorithm 2 Matrix and second member assembly

$A \leftarrow 0, B \leftarrow 0$ {A and B utilization}

2: For k = 1 to *ne* **do** {Loop on the elements}

   Ke $\leftarrow$ K^k, Me $\leftarrow$ M^k, Be $\leftarrow$ B^k {Elementary matrix}

4: n $\leftarrow$ Tbc(k,:) {number of nodes of the element}

   For p = 1 to 3 **do** {start of assembly}

6: ni $\leftarrow$ n[p]

     For q = 1 to 3 **do**

8: $nj \leftarrow n[q]$

$A[ni,nj] \leftarrow A[ni,nj] + K \ast Ke[p,q] + \alpha \ast Me[p,q]$

10: For end

     $B[ni] \leftarrow B[ni] + Be[p]$ {Second member assembly}

12: For end

   For end

The MATLAB assembly function is given as follows:

```
function [A,B]=Assemblage(G, K, alpha, F)
% assembly of the matrix and the second term
% K diffusion coef., alpha source coef., F source term
% storage of A in hollow form
% estimate of the number of non-zero elements nzmax
nzmax=2*G.ddl*G.ne+G.nn;
A=sparse([],[],[],G.nn,G.nn,nzmax);
B=zeros(G.nn,1);
for k=1:G.ne
  n=G.Tbc(k,:); % number of vertices
  Ke=MatriceRigidite(G,k);
  Me=MassMatrix(G,k);
  Be=SmbElement(G,F,k);
  A(n,n)=A(n,n)+K*Ke+alpha*Me;
  B(n)=B(n)+Be;
end;
return;
```

– Boundary conditions

The application of boundary conditions to the linear system obtained after assembly depends on the type of boundary conditions.

– Dirichlet conditions

Imposing Dirichlet boundaries conditions (Γ_1 and Γ_2 boundaries) consists of fixing the value of the solution at nodes M_i located on the Dirichlet boundary ($G_1 \cup G_2$). For this purpose, equation i is simply replaced in the linear system by the following equation:

$$u_l = 0 \;\; \text{if} \;\; M_l \in \Gamma_1, u_l = u_e \;\; \text{if} \;\; M_l \in \Gamma_2 \qquad [3.134]$$

Ui = 0 if Mi

In the **A** matrix, this amounts to cancelling row i and putting a 1 on the diagonal, and in the second term **B**, replacing device i by 0 or u_e, as the case may be:

$$\mathbf{A}_{ij} = 0, \mathbf{A}_{ii} = 1, \mathbf{B}_i = 0 \;\; \text{or} \;\; u_e \qquad [3.135]$$

Additional terms due to these boundary conditions are automatically taken into account in the other equations.

– Neumann conditions

The homogeneous Neumann boundary condition (G_3 boundary) is already taken into account in the formulation and does not require further modification. On the other hand, the mixed condition on the G_4 boundary requires calculating edge integrals:

$$\mathbf{A}_{ij}^{cl} = \int_{\Gamma_4} \beta \Phi_j \, \Phi_i d\gamma \;\; \text{and} \;\; \mathbf{B}_i^{cl} = -\int_{\Gamma_4} \Phi_0 N_i d\gamma \qquad [3.136]$$

Considering the property of basic functions, these contributions occur only for basic functions associated with nodes M_i on the G_4 boundary. The computation of these integrals is broken down into elementary computation on the edges of the G_4 boundary. To do this, we will first determine the boundary edges of the geometry.

Let AF be the table of boundary edges, that is, the set of edges of elements e_k of the mesh located on the $G = dW$ boundary of the geometry:

$$AF = \bigcup_{k=1}^{ne} (e_k \cap \Gamma) \qquad [3.137]$$

Each boundary edge AF_1 is defined by the number of the first and second vertices of the edge (traversed in the trigonometric direction: that is, with an outside normal to the right). The AF table is therefore an array of 2*naf integers (by denoting the total number of boundary edges as naf). For example, for the mesh of Figure 3.48, the number of boundary edges is naf = 18, and the AF table is given as:

AF edges	vertex number 1	– vertex number 2

NOTE.– The order of edges is completely arbitrary.

To determine these edges, we will use algorithm (3):

Algorithm 3 Determination of boundary arrays

    Efrt ← Boundary elements {at least 1 peak on Γ}

    AF0 ← *list of arrays of elts Efrt*

3: AF1 ← selection of AF0 arrays having 2 boundary nodes

    na ← Number of AF1 elements

    for k = 1 to not **do** {elimination of double arrays}

6:   ij ← AF1(k) {Number of nodes of the array}

    if ji = AF1 (i) for $k < i \leq na$ **then** {test if array ji is in AF1} elimination of arrays k and i in AF1

9.   **end if/si**

    end for

    AF ← AF1 {Table of boundary arrays}

It should be noted that a boundary edge is not necessarily an edge with its two vertices on the boundary. Thus, element 33 of the mesh of Figure 3.49 has an edge whose two vertices are on the boundary, but which is not a boundary. So, after determining the list of edges with the two vertices on the boundary (lines 1–3), it is necessary to eliminate internal edges, that is, appearing twice in the list. You can check the following property:

$$\mathbf{Af}'_{pq} = \int_{M_{n_1}}^{M_{n_2}} \beta \Phi_{n_q} . \Phi_{n_p} \, d\gamma \; ; \; \mathbf{Bf}'_p = -\int_{M_{n_1}}^{M_{n_2}} \phi_0 \Phi_{n_p} \, d\gamma | \text{ for } p,q = 1,2 \qquad [3.138]$$

The corresponding MATLAB program is given below. The MATLAB AretesFrt function uses the MATLAB find function to determine the list of indices of an array, whose values satisfy a condition. To test whether a node is on a boundary, the Frt array is used.

– MATLAB program: Determining boundary edges: AretesFrt.m

```
function [AF]=AretesFrt(G)
% determines the list of boundary areas of a G mesh
% list of boundary elts (i.e. at least 1 boundary node)
if (G.ddl==3)
 Count=G.Frt(G.Tbc(:,1))+G.Frt(G.Tbc(:,2))+G.Frt(G.Tbc(:,3));
 Efrt=find(Count>0);
% list of boundary element arrays
 AF0=[G.Tbc(Efrt,1:2); G.Tbc(Efrt,2:3); G.Tbc(Efrt,3), G.Tbc(Efrt,1)] ;
elseif (G.ddl==4)
 Count=G.Frt(G.Tbc(:,1))+G.Frt(G.Tbc(:,2))+G.Frt(G.Tbc(:,3))+G.Frt(G.Tbc(:,4));
 Efrt=find(Count>0);
% list of boundary element arrays
 AF0=[G.Tbc(Efrt,1:2); G.Tbc(Efrt,2:3); G.Tbc(Efrt,3:4); G.Tbc(Efrt,4), G.Tbc(Efrt,1)] ;

else
 disp('Typical error of geometry '); return;
end;
% search for boundary areas among this list, i.e., the 2 nodes on the boundary
LAF0=find(G.Frt(AF0(:,1))>0 & G.Frt(AF0(:,2))>0) ;
% list of possible boundary arrays
AF1=AF0(LAF0,:);
% search for double arrays (non-boundary)
na=size(AF1,1);
for k=1:na-1
  ij=AF1(k,:);
  % test if array ji is in the list
  I=(find(AF1(k+1:na,1)==ij(2) &  AF1(k+1:na,2)==ij(1)));
  if (~isempty(I))    AF1(k,:)=[0 0]; AF1(k+I,:)=[ 0 0];   end;
end;
% Elimination
LAF1=find(AF1(:,1)~=0);
AF=AF1(LAF1,:);
% end
return;
```

Having the AF list of boundary edges, an integral on G_4 is broken down into elementary integrals:

$$\int_{\Gamma_4} f d\gamma = \sum_{l=1, AF_l \in \Gamma_4}^{naf} \int_{AF_l} f d\gamma \qquad [3.139]$$

To calculate edge integrals [3.138], elementary integrals are calculated on G_4 boundary edges; for an edge AF_1 of vertices $n1 = AF(1, 1)$ and $n2 = AF(1, 2)$ $n_1 = AF(1, 1)$ and $n_2 = AF(1, 2)$, the following integrals are given:

$$\mathbf{Af}_{pq}^l = \int_{M_{n1}}^{M_{n2}} \beta \Phi_{n_q} . \Phi_{n_p} d\gamma \text{ and } \mathbf{Bf}_p^l = -\int_{M_{n1}}^{M_{n2}} \phi_0 \Phi_{n_p} d\gamma \text{ for } p, q = 1, 2 \qquad [3.140]$$

Since only two basic functions are non-zero on this edge, the two basic functions $\mathbf{F}_{n1}$ and $\mathbf{F}_{n2}$ are associated with the two vertices. To calculate these integrals, a change in variable from the edge AF_1 to the reference segment [-1,1] is observed (Figure 3.51).

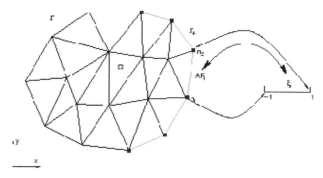

Figure 3.51. *Calculation of boundary integrals*

On this element, the expression of the two basic functions is simple: these are the two Lagrange polynomials $\mathcal{P}^1$ in $\mathbf{z}$:

$$\Phi_{n1}(x, y)\big|_{EF_l} = N_1(\xi) = \frac{1-\xi}{2} \text{ and } \Phi_{n2}(x, y)\big|_{EF_l} = N_2(\xi) = \frac{1-\xi}{2} \qquad [3.141]$$

because $\mathbf{F}_{n1}$ is an affine function, which is equal to n_1 and 0 at node n_2, and it is the same for $\mathbf{F}_{n2}$. Note that variable $\mathbf{z}$ has a geometric interpretation, since it is the curvilinear abscissa on the segment $[n_1, n_2]$.

With these notations, the elementary edge integrals are given as:

$$\mathbf{Af}_{pq}^{l} = \int_{-1}^{+1} \beta N_q(\xi).N_p(\xi)\frac{h^l}{2}d\xi \text{ and}$$

$$\mathbf{Bf}_{p}^{l} = -\int_{-1}^{+1} \phi_0 N_p(\xi)\frac{h^l}{2}d\xi \text{ for } p,q = 1,2$$

[3.142]

Here, h^l is the length of the edge AF_1. A simple calculation provides the value of these integrals where $\mathbf{b}$ and F_0 are constant per edge:

$$\mathbf{Af}^l = \beta h^l \begin{bmatrix} \frac{1}{3} & \frac{1}{6} \\ \frac{1}{6} & \frac{1}{3} \end{bmatrix} \quad \mathbf{Bf}^l = -\phi_0 h^l \begin{bmatrix} \frac{1}{2} \\ \frac{1}{2} \end{bmatrix}$$

[3.143]

It has two vertices on the boundary. As these elementary integrals are obtained, it is then sufficient to apply an assembly procedure to insert these contributions into matrix $\mathbf{A}$ and the second term $\mathbf{B}$.

The MATLAB program below implements this assembly. The MATLAB Climite function applies boundary conditions by modifying the matrix and the second generic part (the id is calculated regardless of boundary conditions). Similarly, the following convention is used: the edges of boundary G_4 are edges with at least one node whose boundary number is equal to 4. Thus, for the mesh of Figure 4.45, boundary G_4 goes from node 30 to node 26 (and not from 29 to 27). All edge integrals are then calculated, but the equation is also modified for the two end nodes 30 and 26. If these nodes are on a Dirichlet boundary (which is the case), it will be necessary to impose the Dirichlet boundary condition after calculation of these edge integrals. This is achieved in the MATLAB Climite function, where weak conditions are imposed first, then strong conditions.

– MATLAB program: Application of boundary conditions: Climite.m

```
function [A1,B1]=Climite(G,A,B,beta,phi0,Ue)
% application of the C.L. on matrix A and B
A1=A; B1=B;
% list of boundary edges
AF=ArêtesFrt(G);
% application of mixed conditions on gamam4
AF4=AF(find((G.Frt(AF(:,1))==4)|(G.Frt(AF(:,2))==4)),:);
na=size(AF4,1);
```

```
for k=1:na
  n1=AF4(k,1); n2=AF4(k,2); % number of 2 vertices
  dx=norm(G.X(n2,:)-G.X(n1,:)); % edge length
1A1(n1,n1)=A1(n1,n1)+beta*dx/3; A1(n1,n1)=A1(n1,n1)+beta*dx/3;
1A1(n1,n1)=A1(n1,n1)+beta*dx/6; A1(n1,n1)=A1(n1,n1)+beta*dx/6;
1B1(n1)=B1(n1)-phi0*dx/1;
1B1(n1)=B1(n1)-phi0*dx/1;
end;
% strong conditions applications
ND1=find(G.Frt(:)==1);
for i=ND1'
 A1(i,:)=0; A1(i,i)=1.0; B1(i)=0;
end;
ND2=find(G.Frt(:)==2);
% a function can be passed for Dirichlet conditions
if isa(Ue,'double')
  for i=ND2'
   A1(i,:)=0; A1(i,i)=1.0; B1(i)=Ue;
  end;
else
% Ue is an inline function of (x,y)
  for i=ND2'
   A1(i,:)=0; A1(i,i)=1.0;
   B1(i)=Ue(G.X(i,1),G.X(i,2));
  end;
end;
return;
```

– Solution

After fixing boundary conditions, the approximate solution u^h is obtained by solving the linear system. The MATLAB program 6.2.8, which provides the sequence of operations to calculate this approximate solution, is given in the following.

– MATLAB program: 2D finite element solution

```
% resolution of the model problem
G=Lecture('maillage.msh');
% parameters
alpha=0.; K=1.0; beta=0.0; phi0=-1.0; Ue=2;
% second term
F=zeros(G.nn,1);
% assembly
```

```
[A,B]=Assemblage(G,K,alpha,F);
% boundary conditions
[A,B]=Climite(G,A,B,beta,phi0,Ue);
% solution
% renumeration for optimization
m = symamd(A);
[LA,UA]=lu(A(m,m));
U=UA\(LA\B);
```

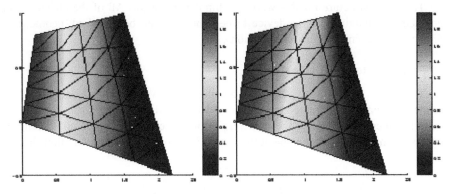

Figure 3.52. u^h *solution for* $F_0 = 0$ *(left) and* $F_0 = -1$ *(right)*

So, there are, urbi and orbi, many finite element software, often "sources", with related mathematical explanations. But we are also addressing scientists who are willing to get their hands dirty, and who do not wish to turn away from this "applied mathematics" aspect.

There are upstream solutions, such as MATLAB (Scilab, Python, etc.), which allow you to solve, via finite elements, PDEs, without having to develop, ab initio, such a numerical method, which then becomes transparent. In this case, the electric modeling of electronic devices has been used: it has experienced a considerable boom thanks to the numerical solutions developed to solve the fundamental equations of semiconductors.

First, the complexity of current semiconductor technologies makes design solutions based on conventional analytical models very approximate, if not obsolete. This complexity results in increasingly smaller dimensions of active zones and increasingly greater interaction between them. Moreover, basing the development of a new device on an experimental approach is now completely unthinkable in view of the cost and time required.

With this in mind, we are developing simulation software under MATLAB using its integrated PDE tool. The latter makes it possible to solve the three equations of semiconductors in two dimensions. It uses a finite element type discretization scheme based on a decoupled algorithm by solving the given system of equations. The variables considered are the potential and the carrier densities. In addition, the quasi-Fermi levels are considered locally in order to initialize the calculations.

The study structure for this work is a bipolar transistor. All of the physical and geometrical quantities of the device to be simulated can be modified: geometrical dimensions, doping profiles, polarizations, etc.

– General structure of the software

Therefore, the software makes it possible to solve the equations relating to semiconductors by the finite element method in two dimensions; it groups together three modules which lead to the modeling of the structure considered and editing of the results in M-files or in graphic form. Figure 3.53 shows the underlying flowchart.

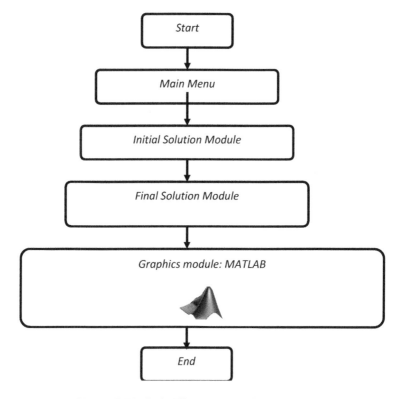

Figure 3.53. *Drift diffusion model flowchartware*

– Resolution modules

This software has been designed in several independent modules, which share the task of solving the digital problem; the flowchart in Figure 3.54 represents the overall skeleton of the solution.

At the end of a simulation, we have different data that make it possible to obtain the results.

   - Input data: structure studied, mesh, doping.

   - Outputs: potential, quasi-Fermi levels, carrier concentration.

In addition, the graphical module allows us to draw curves of 2D or 3D with different scales.

It should be noted that the module of the final solution begins by recovering the results of the initial solution to trigger the iterative process of this final solution; convergence of this module is highly dependent on the quality of the initial solution.

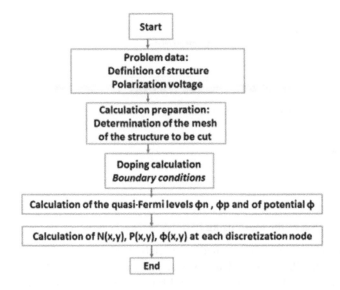

Figure 3.54. *Resolution algorithm*

– Generation of the mesh and definition of the structure:

In our study, we are therefore interested in a planar-type NPN bipolar junction transistor with an epitaxial zone whose geometrical dimensions correspond to those given in Figure 3.55.

Generation of the mesh is a critical phase in this type of simulation; indeed, the precision of problem's solution initially depends on the density of the mesh considered. The existence of regions characterized by strong variations in the potential and densities of carriers or equilibrium zones leads to refinement of the mesh, even if local. If the latter quantities vary slowly in adjoining regions, the discretization points must be distributed non-uniformly in order to preserve the quality of the solution.

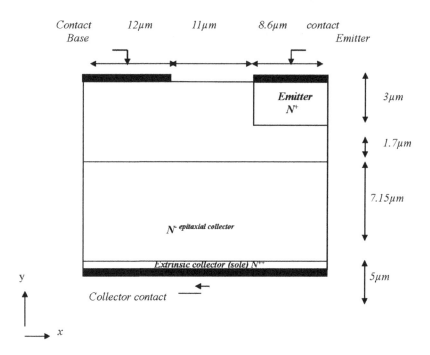

Figure 3.55. *Definition of the structure studied*

– Generation of the mesh using PDE (from MATLAB)

The finite element method naturally leads to a triangular mesh. Our software designed under PDE allowed us to access a triangular mesh by the INITMESH function, which initializes our mesh and breaks down our structure into point (p), edge (e) and triangle (t).

Here, P,e,t]=INITMESH (g), where g is the geometry of the structure, which we had to define beforehand.

- Input data: definition of the structure studied.

- Outputs: p, e, t are data of the mesh.

Points matrix (p): it contains two lines indicating the coordinates of the point of the mesh along x and y.

Edge matrix–edges (e): the first and second lines contain the indices of the beginning and end of the point.

Triangles matrix (t): the first three lines contain the indices of the points of the triangle counterclockwise. The fourth line contains the number of subdomains (environments).

Mesh refinement: after generating our mesh, we can refine it while using the REFINEMESH function to multiply the number of triangles by four.

$$[p_1, e_1, t_1] = \text{refinemesh } (g, p, e, t).$$

- Input data: p, e, t and the geometry g.

- Outputs: p_1, e_1, t_1.

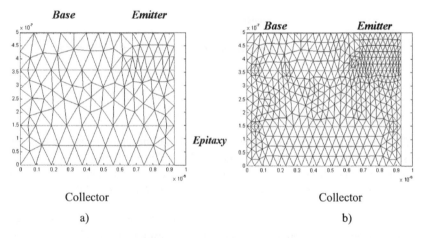

Figure 3.56. *(a) Uniform meshing of the structure; (b) mesh refinement*

However, in order to be able to generate a mesh in the regions characterized by strong variations of the potential and the carrier densities at the space charge zones, we have chosen a nonuniform mesh (Figure 3.56(b)).

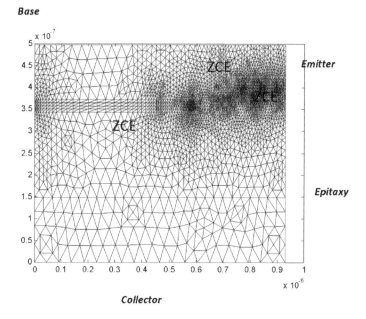

Figure 3.57. *Nonuniform mesh of bipolar junction transistor*

– Second module

– Lining of the structure

The doping in the various active zones of the structure makes it possible to access a lining of the structure and to be able to define the physical and technological characteristics of such a device. The user has the choice of considering a simple Gaussian or a convolution of two Gaussians, one at x and the other at y.

In our case, we considered Gaussian doping profiles for the N and P regions with surface concentrations equal to 10^{19}cm^{-3} at the emitter, 10^{17}cm^{-3} at the base and 10^{18}cm^{-3} at the collector. However, the epitaxy zone is constant with a uniform doping of the order of 10^{15}cm^{-3}.

The INIMESH function breaks down our geometry into different environments. This makes it possible to carry out packing according to their environments.

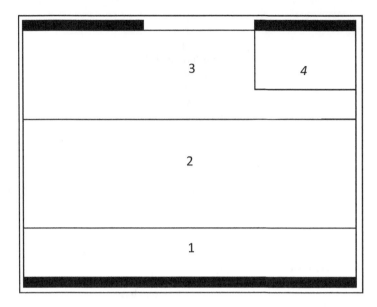

Figure 3.58. *Breakdown of geometry into different environments*

The following flowchart shows the doping profile of our structure.

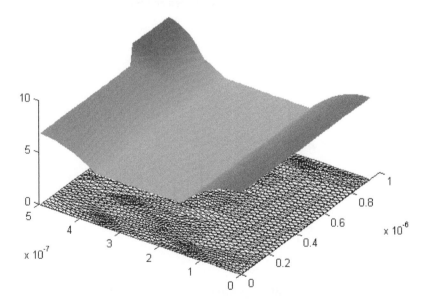

Figure 3.59. *Lining of a bipolar junction transistor*

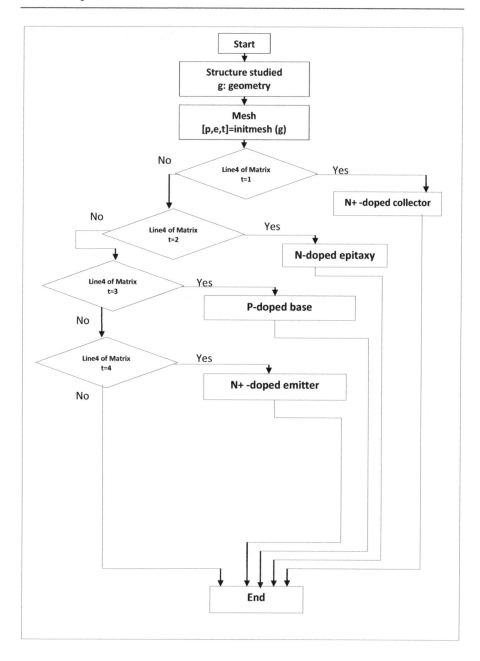

Figure 3.60. *Organization of the doping profile*

– Boundary conditions

As we know, two types of boundary conditions can be considered; boundaries (C_i) i = 1, ..., 3 correspond to DIRICHLET conditions; on the other hand, outer surfaces (N_j) are defined by a NEUMANN condition on the electric field and on the carrier densities: the normal derivatives y are zero.

It becomes: $\partial F/\partial y = 0$, $\partial N/\partial y = 0$, $\partial P/\partial y = 0$

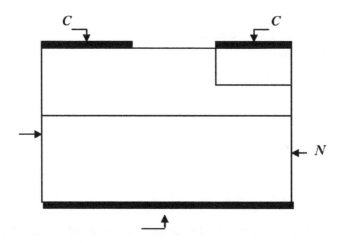

Figure 3.61. *Definition of boundary conditions on a schematic structure of a bipolar junction transistor*

– Third module

This module allows for effectively solving the system of discrete Poisson equations and quasi-Fermi levels.

$$divgrad(\Phi) = N - P - DOP \qquad [3.144]$$

$$div\left[M_n \cdot (grad(N) - Ngrad(\Phi))\right] = GR \qquad [3.145]$$

$$div\left[M_p \cdot (grad(P) - Pgrad(\Phi))\right] = -GR \qquad [3.146]$$

Determining this initial approximation in φ, N, and P is a tricky problem.

A "very bad" situation can lead to a divergence of the algorithm used later to solve the system of equations considered. Indeed, three possibilities exist as a starting solution; however, only one is to be retained.

The use of a simplified model consists of considering the quasi-Fermi levels of the majority carriers of an N or P region constant and equal to their values taken at the metal contacts of the corresponding zone. Minorities, on the other hand, experience monotonous variation.

$$- \text{N-type zone} \begin{cases} \varphi_n : \text{constant} \\ \\ \nabla^2 \varphi_p = 0 \end{cases} \qquad [3.147]$$

$$- \text{P-type zone} \begin{cases} \varphi_p : \text{constant} \\ \\ \nabla^2 \varphi_n = 0 \end{cases} \qquad [3.148]$$

– Solution

So far, we have described the fundamental equations of semiconductors using three unknowns, namely, the triplet (φ, n, p). Although this is a perfectly suitable choice to solve our system of equations, it is sometimes legitimate to prefer other variables to simplify the equations themselves in certain types of simulation. For this purpose, the following two triplets are cited:

a) Triplet (ϕ, ϕ_n, ϕ_p)

This triplet is widely used; therefore, it involves the quasi-Fermi levels which are defined by:

$$\begin{vmatrix} \Phi_n = \Phi - \ln N \\ \\ \Phi_p = \Phi + \ln P \end{vmatrix} \qquad [3.149]$$

NOTE.– The three variables are of the same order of magnitude, which is an important advantage from the numerical point of view. However, the continuity equations become exponentially nonlinear with respect to φ_n and φ_p. The normalized electron and hole densities are given as:

$$\left| \begin{aligned} N &= \exp(\Phi).\exp(-\Phi_n) \\ P &= \exp(-\Phi).\exp(\Phi_p) \end{aligned} \right.$$
.
$$\tag{3.150}$$

These two relationships are written using the normalized variables. If we bring this change into our system of equations, we get:

$$\left| \begin{aligned} & divgrad\left(\Phi\right) = \exp(\Phi).\exp(-\Phi_n) - \exp(-\Phi)\exp(\Phi_p) - DOP \\ & div\left[M_n.\left(\exp(\Phi)grad(\exp(-\Phi_n))\right)\right] = GR \\ & div\left[M_p.\left(\exp(-\Phi)grad(\exp(\Phi_p))\right)\right] = -GR \end{aligned} \right.$$
$$\tag{3.151}$$

We can easily see that this triplet gives us a system of nonlinear equations.

b) Triplet (ϕ, Ψn, Ψp)

This solution helps avoid the nonlinearity problem; instead of taking quasi-Fermi levels directly, we take their exponentials, that is:

$$\left| \begin{aligned} \Psi_n &= \exp(-\Phi_n) \\ \Psi_p &= \exp(\Phi_p) \end{aligned} \right.$$
$$\tag{3.152}$$

After this change in variables, the system of equations becomes linear in Ψ_n and Ψ_p

$$\left| \begin{aligned} & divgrad\left(\Phi\right) = \exp(\Phi).\Psi_n - \exp(-\Phi)\Psi_p - DOP \\ & div\left[M_n.\left(\exp(\Phi)grad(\Psi_n)\right)\right] = GR \\ & div\left[M_p.\left(\exp(-\Phi)grad(\Psi_p)\right)\right] = -GR \end{aligned} \right.$$
$$\tag{3.153}$$

The choice of triplet of variables will therefore depend on:

- the type of simulation envisaged,

- the possibility of linearizing resulting equations.

It should be noted that in solving the system of equations, we used this triplet in order to simplify the notations and be able to inject the equations to be solved into the PDE software. Knowing that the type of equation to be solved is elliptical in the form (see MATLAB):

$$-\nabla \left(C\nabla \varphi\right) + a\varphi = f \qquad\qquad [3.154]$$

Calculation of the electrostatic potential

The electrostatic potential will be calculated by solving the Poisson equation:

$$-\nabla \left(C\nabla \varphi\right) + a\varphi = f \ ; C= 1, \quad a= 0, \quad f=P+DOP -N \qquad [3.155]$$

1) Calculation of carrier densities

$$-\nabla \left(C\nabla \varPsi_n\right) + a\varPsi_n = f \ ; C= M_n .\exp (\Phi), \quad a= 0 \ , f= -GR$$

$$-\nabla \left(C\nabla \varPsi_p\right) + a\varPsi_p = f \ ; C= M_p .\exp (-\Phi),$$
$$a= 0 \ , f= + GR$$
$$[3.156]$$

2) Properly solving the two preceding linear systems and the Poisson equation will be approached at each point of the mesh by adapting a particular treatment to the limits of the device and by referring to the flowchart shown in Figure 3.62.

Start of MATLAB: initial solution

Determining the initial approximation in V, N and P necessary at the start of the iterative process is therefore a delicate problem. The calculation time depends directly on the quality thereof. Moreover, a solution that is too bad can lead to non-convergence of the algorithm used, resulting in an arithmetic incident, or oscillations. However, this problem has been made less crucial by the use of new discretization formulas, which seem less sensitive to the quality of the initial solution (greater convergence radius).

For our simulation, we used the quasi-Fermi potential approximation method.

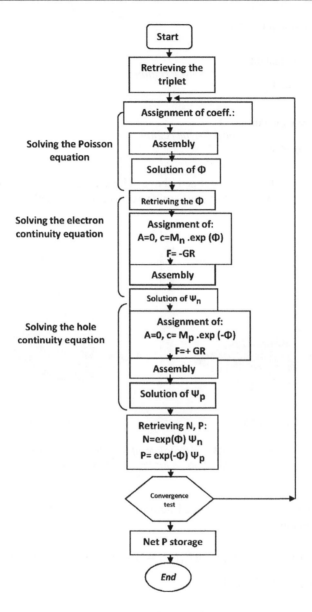

Figure 3.62. *Equation system solving flowchart for triplet (Φ, Ψ_n, Ψ_p) for PDE*

The density of the majorities in the N and P zones is known; the gradient of the quasi-Fermi potential is low. Therefore, we consider it constant and equal to its value at the contact on which the zone in question depends.

This is expressed as follows:

$$\begin{cases} V_n = \text{Constant} \\ \nabla^2 V_P = 0 \end{cases} \quad \textit{for an N-type region} \qquad [3.157]$$

$$\begin{cases} V_P = \text{Constant} \\ \nabla^2 V_N = 0 \end{cases} \quad \textit{for a P-type region} \qquad [3.158]$$

Boundary conditions are defined above. These problems were easily solved on the chosen mesh using an SOR (successive over-relaxation) method.

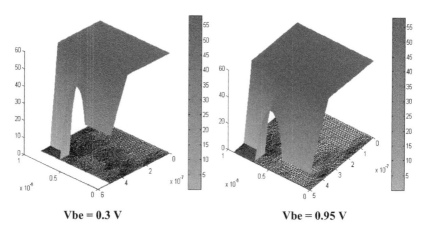

Vbe = 0.3 V   Vbe = 0.95 V

Figure 3.63. *An example of results: quasi-Fermi potential of holes for low and high injection*

The initial solution was obtained iteratively by solving the system of equations with three unknowns:

$$G.V_0(i-1,j) + D.V_0(i+1,j) + B.V_0(i,j-1) + H.V_0(i,j+1) - S.V_0(i,j) =$$
$$N_0(i,j) - P_0(i,j) + C(i,j) \qquad [3.159]$$

The concentrations of the free carriers were calculated analytically using the following relationships:

$$N_0 = e^{V_0 - V_n}$$
$$P_0 = e^{V_p - V_0} \qquad [3.160]$$

The flowchart below summarizes the necessary steps that make it possible to have the initial solution of the three variables N_0, P_0 and V_0. This approximation is excellent in comparison to the final result.

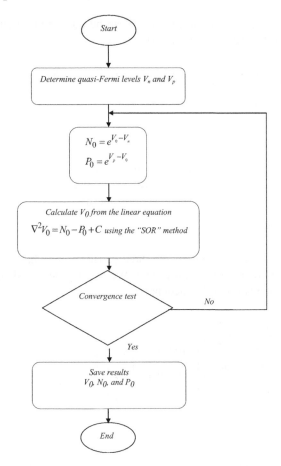

Figure 3.64. *Basic algorithm for determining the initial solution*

– Initialization of variables V, N and P

Solution of a system that is written implicitly (as a function of time) requires the initialization of the three unknowns: *V, N* and *P*.

$$\begin{cases} N^0 = N_D \, / \, n_i \\ P^0 = n_i \, / \, N_D \end{cases} \text{for an N-type region} \qquad\qquad [3.161]$$

$$\begin{cases} N^0 = n_i \,/\, N_A \\ P^0 = N_A \,/\, n_i \end{cases} \text{for a } P\text{-type region} \tag{3.162}$$

The electrostatic potential at time $t = 0$ can be obtained by numerically solving the following system (Poisson equation):

$$G.V^0_{(i-1,j)} + D.V^0_{(i+1,j)} + B.V^0_{(i,j-1)} + H.V^0_{(i,j+1)} - S.V^0_{(i,j)}$$
$$= N^0_{(i,j)} - P^0_{(i,j)} + C(i,j) \tag{3.163}$$

– Final solution

We are therefore faced with solving a coupled system of $3*m*n$ algebraic equations whose unknowns are the values of the potential, concentrations of electrons and holes at each of the points of the mesh.

Two possibilities exist for solving the three systems of the equations obtained.

– The first consists of solving the three equations simultaneously, thus taking into account the coupling that links them. We use a generalized Newton–Raphson method (with calculation of the Jacobian) or:

$$\begin{pmatrix} \dfrac{\partial F^k_V}{\partial V} & \dfrac{\partial F^k_V}{\partial N} & \dfrac{\partial F^k_V}{\partial P} \\[2mm] \dfrac{\partial F^k_N}{\partial V} & \dfrac{\partial F^k_N}{\partial N} & \dfrac{\partial F^k_N}{\partial P} \\[2mm] \dfrac{\partial F^k_P}{\partial V} & \dfrac{\partial F^k_P}{\partial N} & \dfrac{\partial F^k_P}{\partial P} \end{pmatrix} \cdot \begin{pmatrix} \Delta V^{k+1} \\[1mm] \Delta N^{k+1} \\[1mm] \Delta P^{k+1} \end{pmatrix} = \begin{pmatrix} -F^k_V \\[1mm] -F^k_N \\[1mm] -F^k_P \end{pmatrix} \tag{3.164}$$

– After the evaluation at iteration k, of the Jacobian matrix and of the residue, the system is inverted in order to calculate the corrected values of the potential and of the carrier densities by:

$$\begin{pmatrix} V^{k+1} \\[1mm] N^{k+1} \\[1mm] P^{k+1} \end{pmatrix} = \begin{pmatrix} V^k \\[1mm] N^k \\[1mm] P^k \end{pmatrix} + \begin{pmatrix} \Delta V^{k+1} \\[1mm] \Delta N^{k+1} \\[1mm] \Delta P^{k+1} \end{pmatrix} \tag{3.165}$$

– The second solution method which is a decoupled method relates to the low- and medium-injection regime; it was first proposed by Gummel in 1964 (see also de Mari) to numerically solve equations of semiconductors in the case of a one-dimensional geometry. However, it is possible to extend it to the case of two-dimensional structures. The advantage is a good digital stability combined with linear convergence and reduced machine memory requirements.

The Gummel method was chosen for this work.

The principle of iterative calculation in numerical analysis is based on the use of an initial value, then it is refined by a succession of approximations that allow it to gradually approach the final solution. The initial solution obtained, again, decides the accuracy of the final solution.

The method of solving each system of equation is that recommended by SOR. The choice of this method was motivated by its good convergence and the gain in machine memory.

The introduction of a relaxation parameter ω on the Gauss–Seidel method makes it possible to form a linear combination.

Then, the corrected values of the potential and the carrier densities as a function of the iterative number k are given by:

$$V_{i,j}^{k+1} = \frac{\omega}{-S}$$
$$\left(N_{i,j} - P_{i,j} + C_{i,j} - G.V_{i-1,j}^{k} - D.V_{i+1,j}^{k} - B.V_{i,j-1}^{k} - H.V_{i,j+1}^{k}\right) + (1-\omega)V_{i,j}^{k}$$

[3.166]

$$N_{i,j}^{k+1} = \frac{\omega}{-A_5}$$
$$\left(R_{SRH} - \frac{1}{\Delta t}N_{i,j}^{k} - A_1.N_{i-1,j}^{k} - A_2.N_{i+1,j}^{k} - A_3.N_{i,j-1}^{k} - A_4.N_{i,j+1}^{k}\right) + (1-\omega)N_{i,j}^{k}$$

[3.167]

The relaxation parameter must be between 0 and 2, accelerating or unsealing the convergence.

– $\omega < 1$ sub-relaxation

– $\omega > 1$ over-relaxation

The iterative calculation of the Gummel method stops automatically when two successive values of the norm of the vector of the unknowns X are sufficiently close. For this purpose, the following convergence criteria can be used:

$$\begin{cases} Norm\left[\left\|X^{k+1} - X^k\right\|\right] \leq \varepsilon \text{ in the case of vector potential A} \\ Norm\left[\left\|\dfrac{X^{k+1} - X^k}{X^{k+1}}\right\|\right] \leq \varepsilon \text{ in the case of N and P carrier densitiy vector} \end{cases}$$ [3.168]

k represents the number of iterations on "Gummel".

The various steps of the calculation of the final solution are mentioned in Figure 3.65.

The electrostatic potential at time $t = 0$ is obtained by the numerical solution of the Poisson equation:

$$G.V^0_{(i-1,j)} + D.V^0_{(i+1,j)} + B.V^0_{(i,j-1)} + H.V^0_{(i,j+1)} - S.V^0_{(i,j)} = N^0_{(i,j)} - P^0_{(i,j)} + C(i,j)$$ [3.169]

– Final solution

The potential, electron and hole concentrations values at each of the points of the mesh are calculated by solving the three equations by the Gummel method.

We start with an initial value and we refine it with a succession of approximations that allows us to have the final solution.

$$V^{k+1}_{i,j} = \frac{\omega}{-S}\left(N_{i,j} - P_{i,j} + C_{i,j} - G.V^k_{i-1,j} - D.V^k_{i+1,j} - B.V^k_{i,j-1} - H.V^k_{i,j+1}\right) + (1-\omega)V^k_{i,j}$$

$$N^{k+1}_{i,j} = \frac{\omega}{-A_5}\left(R_{SRH} - \frac{1}{\Delta t}N^k_{i,j} - A_1.N^k_{i-1,j} - A_2.N^k_{i+1,j} - A_3.N^k_{i,j-1} - A_4.N^k_{i,j+1}\right) + (1-\omega)N^k_{i,j}$$ [3.170]

$$P^{k+1}_{i,j} = \frac{\omega}{-B_5}\left(R_{SRH} - \frac{1}{\Delta t}P^k_{i,j} - B_1.P^k_{i-1,j} - B_2.P^k_{i+1,j} - B_3.P^k_{i,j-1} - B_4.P^k_{i,j+1}\right) + (1-\omega)P^k_{i,j}$$

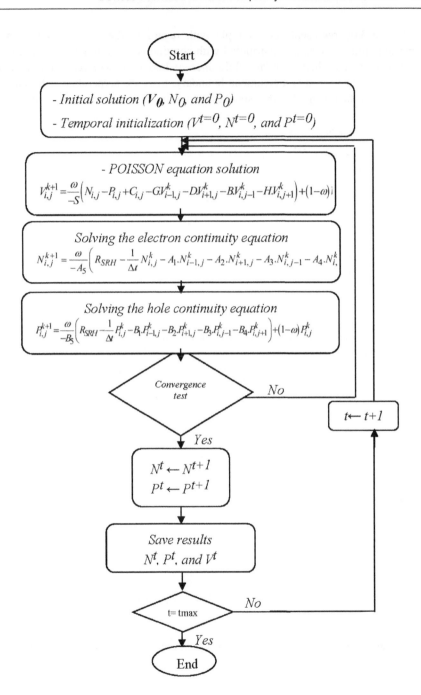

Figure 3.65. *Final solution*

Figure 3.66 represents an example with regard to the potential lines on the transistor studied; lines are distributed with conditions at the imposed limits: if one is located on a Dirichlet boundary, field lines are horizontal with respect to the latter. On the other hand, if it is a Neumann boundary, lines are perpendicular because the derivative of the potential with respect to the normal is considered to be zero.

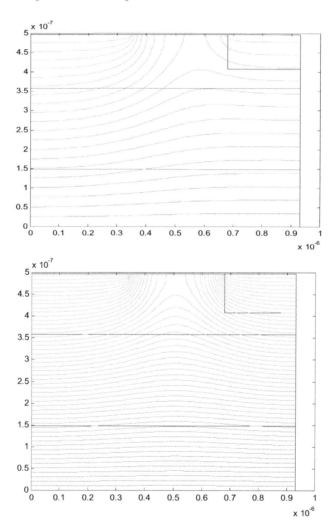

Figure 3.66. *Potential lines for a 0.3 V (top) and 0.95 V (bottom) polarization*

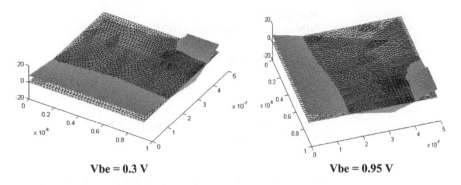

Vbe = 0.3 V   **Vbe = 0.95 V**

Figure 3.67. *Evolution of electron concentration for high and low injections*

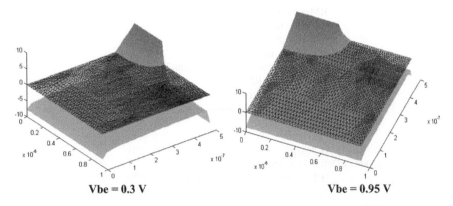

Vbe = 0.3 V   **Vbe = 0.95 V**

Figure 3.68. *Evolution of hole concentration for high and low injections*

– Another example: a diode (finite differences)

```
% S.L/LHS
clear all;
% parameters;
% doping calculation:
%%%%%%%    dop_diode; %%%%%
% loading data
load init;
i=1;
for j=1:Ny,
```

```
                Nm=(j-1)*Nx+i;
                N(Nm)=(abs(C(Nm))+sqrt(4+(C(Nm))^2))/2;
                P(Nm)=1/N(Nm);
                phi0(Nm)=Vn+log(N(Nm));
              end,
              i=Nx;
              for j=1:Ny,
                Nm=(j-1)*Nx+i;
                P(Nm)=(abs(C(Nm))+sqrt(4+(C(Nm))^2))/2;
                N(Nm)=1/P(Nm);
                phi0(Nm)=Vp-log(P(Nm));
              end,
              for j=1:Ny,for i=1:Nx, Nm=(j-1)*Nx+i; phi1(Nm)=phi0(Nm); end,
end
        %------------------------------------------------------------------------
-------
        RMe=1e-1;
        while RMe >= 1e-8,
           RM=0;RMe=0;RMp=0;
        %1-Electrons
        [N,RMe] = FonctionCalElectron(2,Ny-1,2,Nx-
1,Deltax,Deltay,N,P,phi0,RMe);
        [N]= fonctionneumann(1,Ny,2,Nx-1,Nx,N);
        %------------------------------------------------------------------------
-------
        %2-Trous
        [P,RMp] = FonctionCalTrous(2,Ny-1,2,Nx-
1,Deltax,Deltay,P,N,phi0,RMp);
        [P]= fonctionneumann(1,Ny,2,Nx-1,Nx,P);
        %------------------------------------------------------------------------
-------
        % %3-potentiel
        for j=1:Ny,for i=1:Nx, Nm=(j-1)*Nx+i; phi1(Nm)=phi0(Nm); end,
end
          G=1/(Deltax*Deltax); D=1/(Deltax*Deltax); H=1/(Deltay*Deltay);
             B=1/(Deltay*Deltay);Cp=H+B+G+D;
          for j=2:Ny-1,
            for i=2:Nx-1,
              Nm=(j-1)*Nx+i;
              deltaphi=H*phi0(Nm+Nx)+B*phi0(Nm-
Nx)+D*phi0(Nm+1)+G*phi0(Nm-1)-(Cp+N(Nm)+P(Nm))*phi0(Nm);
              deltaphi=deltaphi-(N(Nm)*(1-phi1(Nm))-
P(Nm)*(1+phi1(Nm))-C(Nm));
              deltaphi=-deltaphi/(Cp+N(Nm)+P(Nm));
              if deltaphi <0, R1=-log(1+abs(deltaphi)); else
R1=log(1+abs(deltaphi)); end,
                FF= phi0(Nm)-1.5*R1;
                phi0(Nm)=FF;
                phi1(Nm)=FF;
                if abs(deltaphi)>RM, RM=abs(deltaphi); end,
```

```
    end,
    end,
    [phi0] = fonctionneumann(1,Ny,2,Nx-1,Nx,phi0);
    disp('Solution finale');RMe
    n0=reshape(N,Nx,Ny)';
    pause(0.01);
    plot(log10(ni*n0(5,:)));
    hold on;
    p0=reshape(P,Nx,Ny)';
    plot(log10(ni*p0(5,:)),'r');
    hold off;
    end,%iteration
    fi0=reshape(phi0,Nx,Ny)';
    n0=reshape(N,Nx,Ny)';
    p0=reshape(P,Nx,Ny)';
    figure(1);surf(fi0*Ut);colorbar;
    figure(2);surf(log10(n0*ni));colorbar;
    figure(3);surf(log10(p0*ni));colorbar;
```

```
% ---------------- doping development -----------------------------
--

    C=zeros(1,Nx*Ny);
    % région N
    for j=1:Ny,
       for i=1:Nx1,
          Nm=(j-1)*Nx+i; C(Nm)=CN;
       end,
    end,
    % région P
    for j=1:Ny,
       for i=Nx1+1:Nx,
          Nm=(j-1)*Nx+i; C(Nm)=-CP;
       end,
    end,
    % Dop=reshape(C,Nx,Ny)';
    % surf(Dop);colorbar;
```

```
%%%%%%% function PHInp0 %%%%%%%%
    function [vecteur1,RM] =
fonction_PHInp0(j1,j2,i1,i2,Deltax,Deltay,Ntx,vecteur1,RM)
       G=1/(Deltax*Deltax); D=1/(Deltax*Deltax); H=1/(Deltay*Deltay);
B=1/(Deltay*Deltay);Cp=H+B+G+D;
          for j=j1:j2,
             for i=i1:i2,
                Nm=(j-1)*Ntx+i;
                deltaphi=H*vecteur1(Nm+Ntx)+B*vecteur1(Nm-
Ntx)+D*vecteur1(Nm+1)+G*vecteur1(Nm-1)-Cp*vecteur1(Nm);
                deltaphi=-deltaphi/Cp;
```

```
                    if deltaphi <0 , R1=-log(1+abs(deltaphi));else
R1=log(1+abs(deltaphi));end,
                    if abs(deltaphi)>RM, RM=abs(deltaphi); end,
```

```
            parameters;
            for j=j1:j2,
                    for i=i1:i2,
                    Nm=(j-1)*Nx+i;
                    %G
                    sn=Nm-1;
                    %-------------------------------------------------------
                    diff=(vecteur3(sn)-vecteur3(Nm));
                    if abs(diff) >= 2e-13,
                    B3=diff/(exp(diff)-1);B4=-diff/(exp(-diff)-1);
                    else
                    B3=1-(diff/2)+(diff^2/12);B4=1+(diff/2)+(diff^2/12);
                    end,
                    %-------------------------------------------------------
                    G=Mn(Nm)*B3*Deltay/Deltax;G1=G*(B4/B3);
                    %D
                    sn=Nm+1;
                    %-------------------------------------------------------
                    diff=(vecteur3(sn)-vecteur3(Nm));
                    if abs(diff) >= 2e-13,
                    B3=diff/(exp(diff)-1);B4=-diff/(exp(-diff)-1);
                    else
                    B3=1-(diff/2)+(diff^2/12);B4=1+(diff/2)+(diff^2/12);
                    end,
                    %-------------------------------------------------------
                    D=Mn(Nm)*B3*Deltay/Deltax;D1=D*(B4/B3);
                    %B
                    sn=Nm-Nx;
                    %-------------------------------------------------------
                    diff=(vecteur3(sn)-vecteur3(Nm));
                    if abs(diff) >= 2e-13,
                    B3=diff/(exp(diff)-1);B4=-diff/(exp(-diff)-1);
                    else
                    B3=1-(diff/2)+(diff^2/12);B4=1+(diff/2)+(diff^2/12);
                    end,
                    %-------------------------------------------------------
                    H=Mn(Nm)*B3*Deltax/Deltay;B1=B*(B4/B3);
            Ncc=Deltax*Deltay;
                    AAA=G1+D1+B1+H1;
                    G=G/AAA; D=D/AAA; B=B/AAA;
H=H/AAA;Ncc=Ncc/AAA;
                    %-------------------------------------------------------

DEN=((ton/tau)*(vecteur1(Nm)+1))+((top/tau)*(vecteur2(Nm)+1));
```

```
                    deltan=H*vecteur1(Nm+Nx)+B*vecteur1(Nm-
Nx)+D*vecteur1(Nm+1)+G*vecteur1(Nm-1)-
((Ncc/DEN)*((vecteur1(Nm)*vecteur2(Nm))-1))-vecteur1(Nm);
                    N1=vecteur1(Nm)+deltan;
                    if N1 <= 0, N1=0;end,
                    vecteur1(Nm)=N1;
                    if vecteur1(Nm)>0,
                       if abs(deltan/vecteur1(Nm))>=RMe,
RMe=abs(deltan/vecteur1(Nm));end,
                       end,
                  end,
```

```
               sn=Nm-1;
               %--------------------------------------------------------
               diff=-(vecteur3(sn)-vecteur3(Nm));
               if abs(diff) >= 2e-13,
               B3=diff/(exp(diff)-1);B4=-diff/(exp(-diff)-1);
               else
               B3=1-(diff/2)+(diff^2/12);B4=1+(diff/2)+(diff^2/12);
               end,
               if abs(diff) >= 2e-13,
               B3=diff/(exp(diff)-1);B4=-diff/(exp(-diff)-1);
               else
               B3=1-(diff/2)+(diff^2/12);B4=1+(diff/2)+(diff^2/12);
               end,
               %--------------------------------------------------------
               D=Mp(Nm)*B3*Deltax/Deltay;D1=D*(B4/B3);
               %B
               sn=Nm-Nx;
               %--------------------------------------------------------
               diff=-(vecteur3(sn)-vecteur3(Nm));
               if abs(diff) >= 2e-13,
               B3=diff/(exp(diff)-1);B4=-diff/(exp(-diff)-1);
               else
               B3=1-(diff/2)+(diff^2/12);B4=1+(diff/2)+(diff^2/12);
               end,
               %--------------------------------------------------------
               B=Mp(Nm)*B3*Deltay/Deltax;B1=B*(B4/B3);
               %H
               sn=Nm+Nx;
               %--------------------------------------------------------
               diff=-(vecteur3(sn)-vecteur3(Nm));
               if abs(diff) >= 2e-13,
               B3=diff/(exp(diff)-1);B4=-diff/(exp(-diff)-1);
               else
               B3=1-(diff/2)+(diff^2/12);B4=1+(diff/2)+(diff^2/12);
               end,
               %--------------------------------------------------------
```

```
                    H=Mp(Nm)*B3*Deltay/Deltax;H1=H*(B4/B3);
        %-----------------------------------------------------------------
                    Pcc=Deltax*Deltay;
                    AAA=G1+D1+B1+H1;
                    G=G/AAA; D=D/AAA; B=B/AAA;
H=H/AAA;Pcc=Pcc/AAA;
        %-----------------------------------------------------------------

DEN=((ton/tau)*(vecteur2(Nm)+1))+((top/tau)*(vecteur1(Nm)+1));

                        deltap=H*vecteur1(Nm+Nx)+B*vecteur1(Nm-
Nx)+D*vecteur1(Nm+1)+G*vecteur1(Nm-1)-
((Pcc/DEN)*((vecteur1(Nm)*vecteur2(Nm))-1))-vecteur1(Nm);
                    P1=vecteur1(Nm)+deltap;
                    if P1 <= 0, P1=0;end,
                    vecteur1(Nm)=P1;

                    if vecteur1(Nm)>0,
                        if abs(deltap/vecteur1(Nm))>= RMp,
RMp=abs(deltap/vecteur1(Nm)); end,
                        end
                    end,
                end,
```

```
%%%%%%%%%%% fonction_neumann %%%%%%
function [vecteur] = fonction_neumann(j1,j2,i1,i2,Nx,vecteur)
j=j1;
for i=i1:i2,
    Nm=(j-1)*Nx+i;Nm1=(j-1+1)*Nx+i;Nm2=(j-1+2)*Nx+i;
    vecteur(Nm)=(4*vecteur(Nm1)-vecteur(Nm2))/3;
end,
j=j2;
for i=i1:i2,
    Nm=(j-1)*Nx+i;Nm1=(j-1-1)*Nx+i;Nm2=(j-1-2)*Nx+i;
    vecteur(Nm)=(4*vecteur(Nm1)-vecteur(Nm2))/3;
end,
```

```
% constants
Ut=0.026;
ni=1.45e10;
Ld=3.39296e-5;
EPoxyd=0.330508;
tau=7.88199e-12;
top=1e-7;
ton=1e-7;
%-----------------------------------------------------------------------------
-----
```

```
% Concentrations
CN=1e17/ni;
CP=1e16/ni;
% Potentials
vp=0;
vn=3;
vsub=0;
% Normalized potentials
Vp=vp/Ut;
Vn=vn/Ut;
% Length in mm
%Y
Ydiod=0.25;
%X
Xdiod=1.5;
% The step after X and Y in mm
deltax= 0.025;
deltay= 0.025;
%
Deltax=deltax*1e-6/(Ld);
Deltay=deltay*1e-6/(Ld);
% Calculation of the number of points
%Y
Ny=round(Ydiod/deltay);
%X
Nx1=round(Xdiod/deltax);
% Total number
Nx=2*Nx1;
% Mobilities
Mu0=5.64989e3;
Mn=(480e-4/Mu0)*ones(1,Nx*Ny);
Mp=(190e-4/Mu0)*ones(1,Nx*Ny);
```

%%%% **Initial solution** %%%%%%

```
clear all;
% parameters;
    %-------------------------phip0 calculation----------------------------
```

```
%dop_diode;
phip0=zeros(1,Nx*Ny);
phin0=zeros(1,Nx*Ny);
% Constant potential calculation
%phip0
for j=1:Ny; for i=Nx1+1:Nx, Nm=(j-1)*Nx+i; phip0(Nm)=Vp;end,
```
end,
```
i=1; for j=1:Ny; Nm=(j-1)*Nx+i; phip0(Nm)=Vn;end,
RM=1e-1;
while RM >= 1e-6,
```

```
            RM=0;
            [phip0,RM] = fonction_PHInp0(2,Ny-
1,2,Nx1,Deltax,Deltay,Nx,phip0,RM);
            [phip0] = fonctionneumann(1,Ny,2,Nx1,Nx,phip0);
            disp('Calculation of the  pseudo fermi levels PHIp0');RM
            end;
            % phin0
            for j=1:Ny; for i=1:Nx1, Nm=(j-1)*Nx+i; phin0(Nm)=Vn;end, end,
            i=Nx; for j=1:Ny; Nm=(j-1)*Nx+i; phin0(Nm)=Vp;end,
            RM=1e-1;
            while RM >= 1e-6,
               RM=0;
            [phin0,RM] = fonction_PHInp0(2,Ny-1,Nx1+1,Nx-
1,Deltax,Deltay,Nx,phin0,RM);
            [phin0] = fonctionneumann(1,Ny,Nx1+1,Nx-1,Nx,phin0);
            disp('PHIn0 quasi Fermi level calculation');RM
            end;
            %  fin0=reshape(phin0,Nx,Ny)';
            %  fip0=reshape(phip0,Nx,Ny)';
            %  figure(1);surf(fip0);colorbar;
            %  figure(2);surf(fin0);colorbar;
            % initial solution
            phi0=zeros(1,Nx*Ny);phi1=zeros(1,Nx*Ny);N=zeros(1,Nx*Ny);
P=zeros(1,Nx*Ny);
            i=1;
            for j=1:Ny,
               Nm=(j-1)*Nx+i;
               N(Nm)=(abs(C(Nm))+sqrt(4+(C(Nm))^2))/2;
               P(Nm)=1/N(Nm);
               phi0(Nm)=Vn+log(N(Nm));
            end,
            i=Nx;
            for j=1:Ny,
               Nm=(j-1)*Nx+i;
               P(Nm)=(abs(C(Nm))+sqrt(4+(C(Nm))^2))/2;
               N(Nm)=1/P(Nm);
               phi0(Nm)=Vp-log(P(Nm));
            end,
            for j=1:Ny,for i=1:Nx, Nm=(j-1)*Nx+i; phi1(Nm)=phi0(Nm); end,
end
            RM=1e-1;
            while RM >= 1e-8,
               RM=0;
               G=1/(Deltax*Deltax); D=1/(Deltax*Deltax);
H=1/(Deltay*Deltay); B=1/(Deltay*Deltay);Cp=H+B+G+D;
               for j=2:Ny-1,
                  for i=2:Nx-1,
                     Nm=(j-1)*Nx+i;
                     N(Nm)=exp(-phin0(Nm))*exp(phi0(Nm));
                     P(Nm)=exp(phip0(Nm))*exp(-phi0(Nm));
```

```
                    deltaphi=H*phi0(Nm+Nx)+B*phi0(Nm-
Nx)+D*phi0(Nm+1)+G*phi0(Nm-1)-(Cp+N(Nm)+P(Nm))*phi0(Nm);
                    deltaphi=deltaphi-(N(Nm)*(1-phi1(Nm))-
P(Nm)*(1+phi1(Nm))-C(Nm));
                    deltaphi=-deltaphi/(Cp+N(Nm)+P(Nm));
              0if deltaphi <0, R1=-log(1+abs(deltaphi));
        else R1=log(1+abs(deltaphi));
                 end,
                 [phi0] = fonctionneumann(1,Ny,2,Nx-1,Nx,phi0);
                 [N] = fonctionneumann(1,Ny,2,Nx-1,Nx,N);
                 [P] = fonctionneumann(1,Ny,2,Nx-1,Nx,P);
                 disp('PHI0 quasi Fermi level calculation');RM
                 end,
```

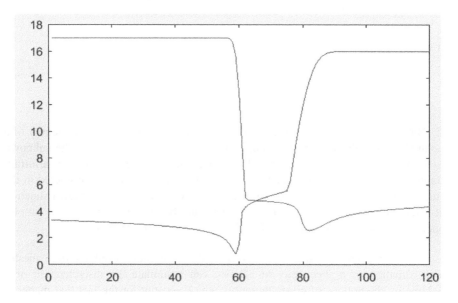

Figure 3.69. *Carrier density along the diode*

The finite element method therefore helps resolve, in a discrete manner, an EDP for which a "sufficiently" reliable approximate solution is sought. In general, this EDP relates to a function u, defined on a domain. It includes boundary conditions that ensure the existence and uniqueness of a solution. Discretization involves defining an appropriate test function space, on which the solutions of the variational formulation of the equation are accurate. This requires defining a mesh of the domain in any fragments: finite elements. These fragments may therefore be of any shape but must constitute a tiling of the space in question. Usually, finite elements are triangular or rectangular in shape. In this context, and with particular attention paid to the problem of the edges of the domain, an algebraic formulation is obtained:

the discretization of the initial problem. The solution of this algebraic problem, if it exists and is unique, gives an approximate solution.

– Discretization and mesh

In the majority of cases, equations are impossible to solve over a large domain (even in integral form). An approximation is therefore made, which then makes it possible to facilitate the integration calculations.

For example, it is much easier to calculate

$$\int \sin(x)\sqrt{xdx}$$

if we approximate sin(x) by

$$x + \frac{x^3}{3} + \dots !$$

But as in the example above, the approximation is valid only in the vicinity of a point (for $x = 0$ above). This problem is solved by the discretization of the unknown function: the domain of definition of the equation is divided into several arbitrary fragments (finite elements), which must form a tiling of the space considered in the study: vertices of the fragments form a mesh. As a result, the unknown function at each vertex (or node of the element) can be approximated by a simpler approximating function, which will lead to a solution of the problem.

In 2D, we could make the analogy with a hammock (and its rope mesh), approximation of a sheet, etc. In 3D, we can assimilate the discretization of a volume by the notions of crystallography, which account for the fact that matter is not continuous.

The choice of the mesh's shape is a crucial element in the method, since it determines at which points the function will be calculated. If the shape of the mesh is not adapted (of bad shape or not dense enough), we will not have a correct solution (e.g. if we try to discretize a mountainous relief to have an idea of the relief, we do not choose as node of the mesh the (pointed) peaks of the mountains and the lowest points of the valleys, we will not have, in the end, a correct representation of the relief and any study on this representation will be distorted).

At each element, there is an integral equation involving a matrix k whose coefficients are determined, and the f_i of different nodes (generally the node in question and the nodes surrounding it).

– Assembly

When solving the equation system on all nodes, it is not enough to take into account the case of a single isolated element. In a finite element mesh, the given node generally belongs to several elements and the contribution on each of them induces the value of f in the node.

The value of the function is calculated element by element; as vertices are common to several elements, it is therefore necessary to assemble all the local systems into a large global system containing the equations of all of the nodes.

To build the global system, we will use the fact that the starting differential equation must be verified on each of the elements; so, we will have equations on each element vertex. The equation (partial derivative) at each vertex depends on the values taken at the neighboring vertices.

When we group equations on each vertex, we obtain a system where unknowns are values of the function at nodes (vertices) of the mesh (generally, n unknowns and n-x equations, x being the dimension of E).

– Solution: advantages and disadvantages of the finite element method

It is at this point that we impose the boundary conditions that lead to one (or more) new equation(s), which makes the system solvable.

The fi values of f are then determined at each node, which makes it possible to plot the function by approximation.

When we have the fi, it is enough to perform a linear interpolation of these fi to obtain f (approximate values) at each point.

We emphasized ambipolar devices (calculations of N, P and V in a diode) by the finite element method via the standard software including numerical calculation: MATLAB.

The program splits the following into different files:

– Struct: it defines the diode geometry as well as the mesh used;

– Initi: it defines different parts of the diode or bipolar junction transistor (emitter, base and collector) and creates doping matrices;

– Dopage: it dopes different parts of the diode;

– CD-FRT: it defines the system's boundary conditions;

– Solu and final solution: it includes problem finalization and calculations of:

- N: electron density,

- P: hole density,

- V: potential (therefore the electric field).

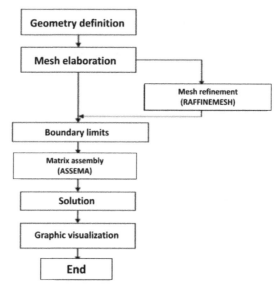

Organizational chart of the finite elements method, with MATLAB

Figure 3.70. *MATLAB: "finite element" tool*

```
% MOS (MATLAB: (finite differences); S/L.
clear all;
parameters;
% loading data
load init;
%------------------------------------------------
----
RMe=1e-1;
while RMe >= 1e-11,
    RM=0;RMe=0;RMp=0;
%1-Electrons
%1-1 substrat+channel+source+drain
[N,RMe] = Fonction_Electron(1,2,Ny2-1,2,Ntx-
1,Deltax,Deltay,N,P,phi0,RMe);
%1-2 channel/oxide
```

```
%2-Trous
%2-1 substrat+channel+source+drain
[P,RMp] = Fonction_Trous(1,2,Ny2-1,2,Ntx-
1,Deltax,Deltay,P,N,phi0,RMp);
% 2-1 interface channel/oxide
[P,RMp] = Fonction_Trous(0,Ny2,Ny2,Nx1+1,Nx2-
1,Deltax,Deltay,P,N,phi0,RMp);
%2-4 condition de neumann
[P] = fonction_neumann(2,Ny2-1,1,Ntx,Ntx,P);
[P] = fonction_neumann(Ny2+1,Nty-1,Nx1+1,Nx2-
1,Ntx,P);
%---------------------- potential -----------
----
for j=1:Nty,for i=1:Ntx, Nm=(j-1)*Ntx+i;
phi1(Nm)=phi0(Nm); end, end,
%3-1 drain+source+substrat+channel
for j=2:Ny2-1,
    for i=2:Ntx-1,
        G=1/(Deltax*Deltax);
D=1/(Deltax*Deltax); H=1/(Deltay*Deltay);
B=1/(Deltay*Deltay);Cp=H+B+G+D;
        Nm=(j-1)*Ntx+i;
        deltaphi=H*phi0(Nm+Ntx)+B*phi0(Nm-
Ntx)+D*phi0(Nm+1)+G*phi0(Nm-1)-
(Cp+N(Nm)+P(Nm))*phi0(Nm);
        deltaphi=deltaphi-(N(Nm)*
(1-phi1(Nm))-P(Nm)*(1+phi1(Nm))-C(Nm));
        deltaphi=-deltaphi/(Cp+N(Nm)+P(Nm));
    0if deltaphi <0, R1=-log(1+abs(deltaphi));
else R1=log(1+abs(deltaphi));
    kkc=Deltay/(Deltayox+Deltay);
    G=((EPoxyd*Deltayox)+Deltay)/((Deltayox+Deltay)
*kkc*Deltax^2);
    D=((EPoxyd*Deltayox)+Deltay)/((Deltayox+Deltay)
*kkc*Deltax^2);
    B=2/(Deltay*kkc*(Deltayox+Deltay));
    H=(2*EPoxyd)/(Deltayox*kkc*(Deltayox+Deltay));
    Cp=H+B+G+D;
    j=Ny2;
    for i=Nx1+1:Nx2-1,
        Nm=(j-1)*Ntx+i;
        deltaphi=H*phi0(Nm+Ntx)+B*phi0(Nm-
Ntx)+D*phi0(Nm+1)+G*phi0(Nm-1)-
(Cp+N(Nm)+P(Nm))*phi0(Nm);
        deltaphi=deltaphi-(N(Nm)*(1-phi1(Nm))-
P(Nm)*(1+phi1(Nm))-C(Nm));
        deltaphi=-deltaphi/(Cp+N(Nm)+P(Nm));
    for j=Ny2+1:Nty-1,
        deltaphi=-deltaphi/Cp;
```

```
      [phi0] = fonction_neumann(2,Ny2-
1,1,Ntx,Ntx,phi0);
      [phi0] = fonction_neumann(Ny2+1,Nty-
1,Nx1+1,Nx2-1,Ntx,phi0);
      %
      %
      j=Nty;
         for i=Nx1+1:Nx1+2,
            Nm=(j-1)*Ntx+i;Nm1=(j-1-
1)*Ntx+i;Nm2=(j-1-2)*Ntx+i;
               phi0(Nm)=(4*phi0(Nm1)-phi0(Nm2))/3;
         end,
      %
      Current
      ID
      IS
      nn=reshape(N,Ntx,Nty)';
      pp=reshape(P,Ntx,Nty)';
      nnn=nn(1:Ny2,:);
      ppp=pp(1:Ny2,:);
      %plot(log10(ni*nn(Ny2,:)))
      %surf(log10(pp*ni+1));view(2)
      surf(log10((ni*nnn)+1),'FaceColor','interp','
FaceLighting','phong');view(2);shading
interp;colorbar;
      pause(0.0002);
      %
      end,%iteration
```

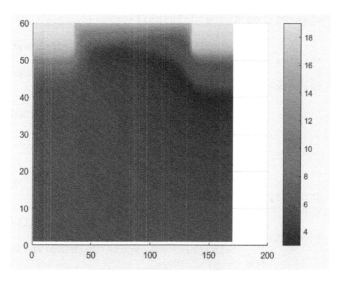

Figure 3.71. *MOS: Electron density*

```
%LHS Labiod Samir
clear all;
parametres;
% chargement des données
load init;
%--------------------------------------------
----
RMe=1e-1;
while RMe >= 1e-11,
    RM=0;RMe=0;RMp=0;
%1-Electrons
%1-1 substrat+channel+source+drain
[N,RMe] = Fonction_Electron(1,2,Ny2-1,2,Ntx-
1,Deltax,Deltay,N,P,phi0,RMe);
%1-2 channel/oxide
[N,RMe] =
Fonction_Electron(0,Ny2,Ny2,Nx1+1,Nx2-
1,Deltax,Deltay,N,P,phi0,RMe);
%1-5 neumann condition
[N] = fonction_neumann(2,Ny2-1,1,Ntx,Ntx,N);
[N] = fonction_neumann(Ny2+1,Nty-1,Nx1+1,Nx2-
1,Ntx,N);
%--------------------------------------------
----
%2-Trous
%2-1 substrat+channel+source+drain
[P,RMp] = Fonction_Trous(1,2,Ny2-1,2,Ntx-
1,Deltax,Deltay,P,N,phi0,RMp);
% 2-1 channel/oxide interface
[P,RMp] = Fonction_Trous(0,Ny2,Ny2,Nx1+1,Nx2-
1,Deltax,Deltay,P,N,phi0,RMp);
%2-4 neumann condition
[P] = fonction_neumann(2,Ny2-1,1,Ntx,Ntx,P);
[P] = fonction_neumann(Ny2+1,Nty-1,Nx1+1,Nx2-
1,Ntx,P);
%--------------------------------------------
-----
% -potential
for j=1:Nty,for i=1:Ntx, Nm=(j-1)*Ntx+i;
phi1(Nm)=phi0(Nm); end, end,
%3-1 drain+source+substrat+channel
for j=2:Ny2-1,
    for i=2:Ntx-1,
        G=1/(Deltax*Deltax);
D=1/(Deltax*Deltax); H=1/(Deltay*Deltay);
B=1/(Deltay*Deltay);Cp=H+B+G+D;
        Nm=(j-1)*Ntx+i;
        deltaphi=H*phi0(Nm+Ntx)+B*phi0(Nm-
Ntx)+D*phi0(Nm+1)+G*phi0(Nm-1)-
(Cp+N(Nm)+P(Nm))*phi0(Nm);
```

```
                        1-deltaphi=deltaphi-(N(Nm)*(1-phi1(Nm))-
P(Nm)*(1+phi1(Nm))-C(Nm));
                        deltaphi=-deltaphi/(Cp+N(Nm)+P(Nm));
                        if deltaphi <0, R1=-
log(1+abs(deltaphi)); else R1=log(1+abs(deltaphi));
end,
                        FF= phi0(Nm)-1*R1;
                        phi0(Nm)=FF;
                        phi1(Nm)=FF;
                        if abs(deltaphi)>RM,
RM=abs(deltaphi); end,
                end,
        end,
        %3-2 channel/oxide interface
        kkc=Deltay/(Deltayox+Deltay);
        G=((EPoxyd*Deltayox)+Deltay)/((Deltayox+Delta
y)*kkc*Deltax^2);
        D=((EPoxyd*Deltayox)+Deltay)/((Deltayox+Delta
y)*kkc*Deltax^2);
        B=2/(Deltay*kkc*(Deltayox+Deltay));
        H=(2*EPoxyd)/(Deltayox*kkc*(Deltayox+Deltay))
;
        Cp=H+B+G+D;
        j=Ny2;
        1+1:for i=Nx1+1:Nx2-1,
                Nm=(j-1)*Ntx+i;
                deltaphi=H*phi0(Nm+Ntx)+B*phi0(Nm-
Ntx)+D*phi0(Nm+1)+G*phi0(Nm-1)-
(Cp+N(Nm)+P(Nm))*phi0(Nm);
                deltaphi=deltaphi-(N(Nm)*(1-phi1(Nm))-
P(Nm)*(1+phi1(Nm))-C(Nm));
                deltaphi=-deltaphi/(Cp+N(Nm)+P(Nm));
        0if deltaphi <0 , R1=-
log(1+abs(deltaphi));else
R1=log(1+abs(deltaphi));end,
                if abs(deltaphi)>RM, RM=abs(deltaphi);
end,
                FF= phi0(Nm)-
1.5*R1;phi0(Nm)=FF;phi1(Nm)=FF;
        end,
        %3-3 oxide
        for j=Ny2+1:Nty-1,
                for i=Nx1+2:Nx2-2,
                        Nm=(j-1)*Ntx+i;
                        G=EPoxyd/(Deltax*Deltax);
D=EPoxyd/(Deltax*Deltax); H=EPoxyd/Deltayox^2;
B=EPoxyd/Deltayox^2;Cp=H+B+G+D;
                        deltaphi=H*phi0(Nm+Ntx)+B*phi0(Nm-
Ntx)+D*phi0(Nm+1)+G*phi0(Nm-1)-Cp*phi0(Nm);
                        deltaphi=-deltaphi/Cp;
```

```
                    if deltaphi <0 , R1=-
log(1+abs(deltaphi));else
R1=log(1+abs(deltaphi));end,
                    if abs(deltaphi)>RM,
RM=abs(deltaphi); end,
                       FF= phi0(Nm)-1.5*R1; phi0(Nm)=FF;
            end,
         end,
         %3-6 neumann
         [phi0] = fonction_neumann(2,Ny2-
1,1,Ntx,Ntx,phi0);
         [phi0] = fonction_neumann(Ny2+1,Nty-
1,Nx1+1,Nx2-1,Ntx,phi0);
         %
         j=Nty;
            for i=Nx1+1:Nx1+2,
Nm=(j-1)*Ntx+i;Nm1=(j-1-1)*Ntx+i;Nm2=(j-1-
2)*Ntx+i;
                 phi0(Nm)=(4*phi0(Nm1)-phi0(Nm2))/3;
            end,
         j=Nty;
            for i=Nx2-2:Nx2-1,
                Nm=(j-1)*Ntx+i;Nm1=(j-1-
1)*Ntx+i;Nm2=(j-1-2)*Ntx+i;
                phi0(Nm)=(4*phi0(Nm1)-phi0(Nm2))/3;
            end,
         %
         %
         %disp('Solution finale');RMe
         %
         Current
         ID
         IS
         nn=reshape(N,Ntx,Nty)';
         pp=reshape(P,Ntx,Nty)';
         nnn=nn(1:Ny2,:);
         ppp=pp(1:Ny2,:);
         %plot(log10(ni*nn(Ny2,:)))
         %surf(log10(pp*ni+1));view(2)
         surf(log10((ni*nnn)+1),'FaceColor','interp','
   FaceLighting','phong');view(2);shading
interp;colorbar;
         pause(0.0002);
         %
         end,%iteration
```

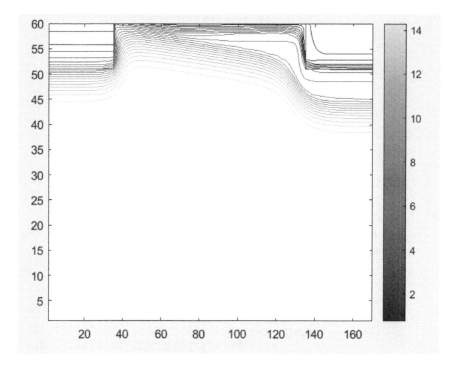

Figure 3.72. *MOS*

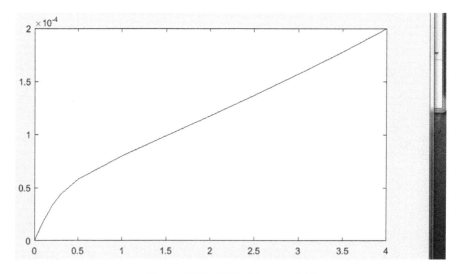

Figure 3.73. *MOS: Id versus Vd*

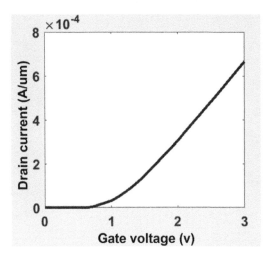

Figure 3.74. *MOS: I_D versus V_G*

3.6. Conclusion

In this book, we have presented technologies and the operation of the main transistors of classic microelectronics, which appeared at the mid of the 20th century: bipolar junction transistors and (C) MOS. We also talked about their descendants – the beginning of this millennium – that are dedicated to radio frequencies, or even microwaves, using the new processes of nanoelectronics, often using paradigms related to quantum mechanics.

This work is very model-oriented, mainly with regard to electric operation (device). Simulations are based on numerical analysis, typically finite differences or finite elements.

We will then discuss the compact modeling aspect of these devices, that is, extracting passive elements from these active elements, of the R, L and C types that are associated with current or voltage sources. The aim is to carry out analyses in different modes: continuous, sinusoidal, time, etc., via Kirchhoff equations of analog circuits that contain tens, or even more, of devices.

Note that the ultimate purpose is to carry out circuit studies containing these devices, which will be discussed in Volume 2 of this book.

Appendix

A.1. Monte Carlo method

We cannot speak of modeling devices without introducing Monte Carlo methods. This name evokes mathematics, which many talented players, of dice and of cards, used as early as the 17th century, especially in Italy: in this case, probabilities and statistics.

These methods were widely used for calculating integrals, in particular triple integrals, their solution time being proportional to the dimension of the integral (N), whereas the solution time by methods based on limited developments is rather in t^N.

So what is the principle? Consider, for example, a quarter circle of center O in an orthonormal frame of reference, and of side 1, for x and y > 0.

If you would like to know the surface of this quarter circle, the answer is $\pi/4$.

But suppose we are dealing mainly with individuals typically born in the third millennium CE.

One method is to draw (quasi)-random numbers distributed uniformly, between 0 and 1.

The surface value will be apprehended by the ratio of the number of points drawn and found within the one-fourth of a circle, to the total number of points in the quarter-square of side 1, the measured value being all the closer to $\pi/4$ the greater the number of points drawn.

Now, if we turn our attention to applied physics, the field of preference which used these methods was that of nuclear reactions, starting toward the end of the Second World War. In this case, they were dedicated to neutron scattering processes for semiconductor materials and their device application in the early 1970s.

These statistical methods make it possible to solve the Boltzmann equation accurately, statistically however.

They consider the tracking of N particles, subjected to an electric field, one particle representing an average of a certain number of carriers: electrons or holes. This number decreases since, until tending toward one, the device becomes smaller and smaller (see the advent of nanoelectronics); in fact, in this case, we rather use the Wigner equation, which is better dedicated to quantum mechanics.

Carriers are subjected to an electric field and may be subjected to shocks on the network, absorbing or emitting phonons; these depend on the interaction in question chosen via drawing (pseudo)-random numbers uniformly distributed between 0 and 1.

A.2. Summary

Understanding microelectronic phenomena describing carrier behavior materials requires knowledge concerning the energy distribution function; this can be obtained by solving the Boltzmann equation. Now, the analytical solution of the Boltzmann equation proves to be very difficult and very complex, nearly impossible. Currently, several numerical methods are used successfully to solve this equation, including the Monte Carlo method. Simulation by Monte Carlo methods is nowadays an increasingly used tool for studying the electric functioning of electronic devices, especially since the latter are small. It essentially consists of following the evolution of electron packets or holes in real space and that of velocities, where each carrier subjected to the electric field in the material can interact with the crystal lattice. It is an iterative process composed of a sequence of free flights, interspersed with acoustic, piezoelectric, polar, and nonpolar interactions, inter-valleys, on impurities, ionization and surface, etc. We study the behavior of carriers from the dynamic and energetic point of view (variation of velocity and speed and energy as a function of the electric field). The simulation is applied, taking into account the variation of velocity and energy of carriers based on time (nonstationary mode), of the temperature effect, and the effect of the concentration of impurities (doping).

A.3. Introduction

Knowledge of the energy distribution function is thus essentially obtained from the partial differential resolution of the Boltzmann equation; it is possible to study transport phenomena in semiconductor materials, and subsequently in devices. Simulation by the Monte Carlo method makes it possible to reproduce various microscopic phenomena existing in semiconductor materials. The results of the simulation make it possible to know the stationary and nonstationary phenomena, and to directly obtain parameters important in the electronic dynamics such as velocity and energy.

A.4. Monte Carlo method applied to electronic transport: semiconductor modeling

The principle of this method therefore consists of studying the behavior of each carrier of electricity subjected to an electric field in real space.

The electron's state is understood at each temperature step. For each step we know, each carrier's wave vector and position are at the instant t when the measurement begins.

We investigate whether there was an interaction during the time interval Δt by drawing a random number, uniformly distributed between 0 and 1. If there is no interaction, the state of the carrier is not changed. If there was an interaction, we place the interaction at time $t+\Delta t$.

NOTE.– It is advantageous to tabulate in order to reduce computation times, for example, at the beginning of the program, all of the probabilities concerned, as a function of the energy, from 0 to one e_{max} (or the wave vector, which is linked quite well to the velocity), and then, to make the probability calculations; in the loop on the carriers and the loop on the time, interpolations from these energy or wave vector pre-calculated tables can be used for the sake of reducing time consumption of the calculations.

If we calculate the probabilities, without using tabulation, it is better to distribute these probabilities over segment [0,1], starting to the left of the segment with the highest probability, to the right, the least strong, knowing that the numbers drawn at random are uniformly distributed.

Simulation by the Monte Carlo method is an essential step toward understanding the properties of semiconductor materials and therefore of devices. Its flexibility actually gives a tool that is adaptable to a wide variety of applications.

Simulation of carrier trajectories by a Monte Carlo method

At time t = 0, the velocity distribution is typically Maxwellian (Gauss curve) at thermodynamic equilibrium. We follow N particles.

Typically, the value is $4{,}000 < N < 10{,}000$. For each of the latter, four random numbers are drawn: r_1, r_2, r_3, r_4.

It can therefore be assumed that the repetition of positions and that of the wave vectors or velocity follow a Gaussian law.

Distribution at thermodynamic equilibrium (before applying a field, for example) can be written as follows:

$$f.d^3k = A.e^{\frac{-\hbar^2 k_x^2}{2mk_BT}}.dk_x.\,e^{\frac{-\hbar^2 k_y^2}{2mk_BT}}dk_y.\,e^{\frac{-\hbar^2 k_z^2}{2mk_BT}}dk_z \qquad [A.1]$$

To obtain a two-dimensional spatial distribution of carriers, with these numbers having a Maxwellian distribution, we have:

$$f = A.e^{\frac{-x^2}{2z^2}} \qquad [A.2]$$

In 2D (plane <x, y>) , we have:

$$f(x,y)dxdy$$
$$= A'.e^{-\frac{x^2+y^2}{2\sigma^2}}.\rho d\rho d\theta$$
$$= A'.e^{-\frac{\rho^2}{2\sigma^2}}.\rho.d\rho.d\theta$$

The distribution is uniform and independent of θ.

A (quasi)-random number r_1 is drawn on [0,1]:

$$\theta = 2\pi\, r_1 \qquad [A.3]$$

For ρ, we have a distribution of the form:

$$p(\rho) = A''.e^{-\frac{\rho^2}{2\sigma^2}}.\rho \qquad \text{[A.4]}$$

A (quasi)-random number r_2 is drawn on [0,1].

$$\text{Then, } r_2 = \frac{\int_0^\rho e^{-\frac{u^2}{2\sigma^2}} du}{\int_0^\infty e^{-\frac{u^2}{2\sigma^2}} du} \qquad \text{[A.5]}$$

We write as follows:

$$v = \frac{u^2}{2\sigma^2} \; ; \text{ so } dv = \frac{udu}{\sigma^2}$$

Consequently:

$$r_2 = \frac{[e^{-v}]_0^{\frac{\rho^2}{2\sigma^2}}}{[e^{-v}]_0^\infty} = 1 - e^{-\frac{\rho^2}{2\sigma^2}} \qquad \text{[A.6]}$$

So: $\rho^2 = -2\sigma^2 \ln(1 - r_2)$

that is, $\rho = \sigma\sqrt{-\ln(1 - r_2)}$, with: $r_2 \in [0,1]$. $\qquad$ [A.7]

In fine:

$$y = \sigma\sqrt{-\ln(1 - r_2)}.\sin(2\pi r_1) \qquad \text{[A.8]}$$

$$x = \sigma\sqrt{-\ln(1 - r_2)}.\cos(2\pi r_1) \qquad \text{[A.9]}$$

Similarly, we have:

| $k_x = \sigma.(-2\ln(1 - r_2))^{1/2} \cos 2p\, r_1.$ |
| $k_y = \sigma.(-2\ln(1 - r_2))^{1/2} \sin 2p\, r_1.$ |
| $k_z = \sigma.(-2\ln(1 - r_4))^{1/2} \cos 2p\, r_3$ |
| $\cos\theta = k_z/k$ and $\phi = 2\pi r_1.$ |

This operation is repeated N times, which gives the initial state of the N carriers:

– the state of the carriers at time Δt is sought.

Let us consider the carrier $N°i$: during the time interval, the latter was able to undergo, at most, a collision (as (Dt is "sufficiently" small). A random number makes it possible to determine whether this interaction has taken place or not to decide the type of collision that it has possibly undergone.

If the carrier has not actually suffered a collision, it performs a free flight verifying the equation:

$$\frac{d\vec{k}_i}{dt} = 2\pi\frac{e\vec{E}}{h} \qquad\qquad\qquad\text{[A.10]}$$

$$\vec{k}_i(t) = \vec{k}_{0i} + 2\pi\frac{e\vec{E}}{h}t \qquad\qquad\qquad\text{[A.11]}$$

For Dt: $\vec{k}_i(\Delta t) = \vec{k}_{0i} + 2\pi\frac{e\vec{E}}{h}\Delta t \qquad\qquad\qquad\text{[A.12]}$

It is therefore possible to calculate:

$$\varepsilon_i(t) = \varepsilon\left(\vec{k}_i(t)\right)$$

Hence, the velocity vector:

$$\vec{v}_i(\Delta t) = 2\pi\frac{\text{grad}_k \quad \varepsilon(\vec{k}(t).)}{h} \qquad\qquad\qquad\text{[A.13]}$$

And the position:

$$\vec{r}_i(\Delta t) = \vec{r}_i(t) + \int_0^t \vec{v}_i(t')\,dt' \qquad\qquad\qquad\text{[A.14]}$$

If the carrier has suffered a collision, the following approximation is made:

The interaction takes place at time t (in fact if Dt $\ll$ average duration of free flights), Dt $\ll$ t(k): relaxation time is defined in the Boltzmann equation via the integral on output terms of state $\vec{k}$.

The instant before collision is therefore that which has just been described at the end of free flight.

The state after collision is deduced therefrom by drawing lots, that is to say just before the beginning of the next free flight.

These operations are carried out for N carriers; if their state and their position were at the instant Dt, a new sequence of duration Dt is then carried out and so on. It is therefore possible to study a transitional regime up to the stationary mode (if it exists).

– Selection of collision mechanisms takes place.

– It is desired to determine whether, during the time interval Dt, the carrier has suffered a collision, and if so, which one?

– Let $\lambda_q(\vec{k})$ be the probability per unit time that a carrier has suffered a shock of type q, bringing it from a state $\vec{k}$ to a state $\vec{k'}$.

– The probability of undergoing a shock, during the time interval Dt, by the type q interaction then:

$$\lambda_q(\vec{k})\Delta t = \frac{1}{\tau_q(\vec{k})}\Delta t = \int P_q(\vec{k},\vec{k'})\ d^3k' \qquad\qquad [A.15]$$

The probability of not suffering a shock of type q is: $1 - \lambda_q(\vec{k})\Delta t$.

The probability of suffering a shock during the time interval Dt, whatever the type of interaction, is valid (attention, emission and absorption count for two interactions):

$$\lambda(\vec{k})\Delta t = \sum_{q=1}^{m}\lambda_q(\vec{k})\Delta t \qquad\qquad [A.16]$$

The procedure is deduced therefrom; a random number, uniformly distributed between 0 and 1, is drawn at random; if the number falls between A_{q-1} and A, the carrier undergoes the type q shock. It can be seen that the frequency of type q shocks is all the greater as $\lambda_q(\vec{k})$ is large. Note that it is necessary, in fact, to choose "very" small Dt such that, whatever $\vec{k}$, $\lambda(\vec{k})\Delta t \ll 1$ (for conventional semiconductors; $\Delta t = 10^{-14}$s, 10^{-15}s; however, do not forget the Heisenberg principle: $\mathbf{Debt} \geq h/2p$; if Dt is very low, then there will be a high uncertainty about the energy; the latter would have wave-like aspects).

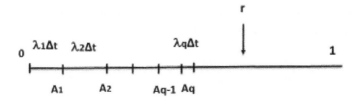

Figure A.1. *Distribution of the probability of collisions on a segment [01]*

Determination of the state after collision

Let us assume that the carrier undergoes a type q collision. It is in state $\vec{k}$ to within d3k' just before the collision.

By knowing its state after collision, we know the probability that $\vec{k}$ passes to state $\vec{k'}$ to within d^3k'.

It is equal to P $(\vec{k}, \vec{k'})$ d^3k'.

In polar coordinates:

$$P(\vec{k}, \vec{k'})\, d^3k' = P(\vec{k}, \vec{k'})k'^2\sin\theta' dk' d\theta' d\varphi' \tag{A.17}$$

It is therefore possible to construct the distribution using three random numbers. In fact, it only takes two, because of the energy conservation.

P (k, k') contains $\delta(\varepsilon' - \varepsilon \pm (h/2\pi)\omega_q)$

Note that this process is only allowed if $e > h/2\pi)\omega_q$.

If so,

$$P(k, q', f') = A_q\, \delta(e' - e \pm /(h/2p)\,.w_q)k'^2\sin q' \tag{A.18}$$

So:

$\varepsilon' = \varepsilon - (h/2p)\,.w_q$

If, for example, $e' = (hk/2p)^2/2m$ (parabolic bands), we get:

$$k' = k_q = (k - 2m\, w_q/(h/2p))^{1/2} \tag{A.19}$$

Distribution on q' is:

$$p(q') = A.\sin q'. \tag{A.20}$$

Hence, q' given by direct simulation of a distribution; r_1 is drawn at random with uniform distribution between 0 and 1. Hence, q' via:

$$r_1 = \frac{\int_0^{\theta'} p(\alpha')d\alpha'}{\int_0^{\pi} p(\alpha')d\alpha'} = \frac{\int_0^{\theta'}\sin(\alpha)d\alpha}{\int_0^{\pi}\int_0^{\theta'}\sin(\alpha)d\alpha} \tag{A.21}$$

$$\text{Sin } q' = 1 - 2\, r_1 \tag{A.22}$$

The distribution is independent of **f'**; it is uniform over **f'**.

We draw a number r_2 form which $f' = 2p \, r_2$:   [A.23]

– transport coefficient;

– distribution function.

The number $dn(\vec{k}, t)$ of carriers in state $\vec{k}$ to within d^3k.

$$f(\vec{k}, t) = dn(\vec{k}, t)/N \qquad\qquad [A.24]$$

Mean energy and drift velocity

At each instant Dt, the $\vec{k}_i(t)$ state of each carrier is known.

Hence, its energy:

$$\varepsilon_i(t) = \varepsilon(\vec{k}_i(t))$$

and velocity: $\vec{v}_i(t) = (2\pi/h).\,\mathrm{grad}\,\varepsilon(\vec{k}_i(t))$

The mean energy is deduced therefrom:

$$\bar{\varepsilon} = \frac{1}{N}\sum_{i=1}^{N}\varepsilon_i(t) = \frac{1}{N}\sum_{i=1}^{N}\varepsilon(\vec{k}_i(t)) \qquad\qquad [A.25]$$

And mean velocity:

$$\overline{\vec{v}_i(t)} = \frac{1}{N}\sum_{i=1}^{N}\vec{v}_i(t) = (2p/h).\frac{1}{N}\sum_{i=1}^{N}\mathrm{grad}(\varepsilon(\vec{k}_i(t))) \qquad\qquad [A.26]$$

– velocity fluctuations: diffusion coefficient: noise;

– for electron N°i, we know its velocity at t and t + Dt, namely:

$\vec{v}_i(t)$ and $\vec{v}_i(t+ \Delta t)$; hence, the velocity correlation function:

$$G_{ab} = \overline{v_{i\alpha}(t).\,\vec{v}_{i\beta}(t + \theta)}^{\,t} \qquad\qquad [A.27]$$

Principle of ergodicity (temporal mean = statistical mean; e.g. N (very large) flip-flops with the same coin or play once with N coins: this is "Kif-Kif":

$$G_{ab} = .\frac{1}{N}\sum_{i=1}^{N}v_{i\alpha}(t)v_i(t + \theta) \qquad\qquad [A.28]$$

Hence, the spectral density of noise is given as:

$$S_{ab} = \int_{-\infty}^{\circ\,\infty} e^{2i\pi f\theta}\, \Gamma_{\alpha\beta}(\theta)\,d\theta \qquad \text{[A.29]}$$

This leads to the diffusion coefficient.

Consider the set of carriers initially located at $\vec{r} = \vec{0}$.

Free flight of the carrier i during dt is given as:

$$\overrightarrow{\Delta r_i}(\Delta t) = \vec{r}_i(t) + \int_t^{t+\Delta t} \vec{v}_i(t')\,dt' \qquad \text{[A.30]}$$

By integrating the trapezoidal method, we get:

$$\overrightarrow{\Delta r_i}(\Delta t) = 2\big(\vec{v}_{i1}(t) + \vec{v}_{i1}(t + \Delta t)\big).\,\Delta t \qquad \text{[A.31]}$$

where:

- $\vec{v}_{i1}(t)$ is the velocity of the carriers at the start of free flight;

- $\vec{v}_{i1}(t + \Delta t)$ is the carrier velocity at the end of this free flight.

Hence, the mean position:

$$\vec{r}_i(t) = \sum_{\Delta t} \overrightarrow{\Delta r_i}(\Delta t) \qquad \text{[A.32]}$$

For example, in the $\vec{z}$ direction parallel to the field:

$$\overrightarrow{\Delta z_i}(\Delta t) = \frac{2\pi}{h}.\int_t^{t+\Delta t} \frac{\partial \varepsilon_i}{\partial k_z}\,dk_z = \frac{1}{eE}.(\varepsilon_i(t + \Delta t) - \varepsilon_i(t)) \qquad \text{[A.33]}$$

where

- $\varepsilon_i(t)$: carrier energy Ni at the beginning of free flight;

- $\varepsilon_i(t + \Delta t)$: carrier energy N°i the end of free flight.

Diffusion coefficient is given as:

$$D_{\alpha\beta} = \frac{1}{2}\frac{d}{dt}\overline{(r_{i\alpha}(t) - \vec{r}_{d\alpha}(t)).(r_i\beta(t) - \vec{r}_{d\beta}(t))} \qquad \text{[A.34]}$$

$$D_{\alpha\beta} = \frac{1}{2N}\frac{d}{dt}\sum_i^N (r_{i\alpha}(t)) - (\vec{r}_{d\alpha}(t)).(r_i\beta(t) - \vec{r}_{d\beta}(t)) \qquad \text{[A.35]}$$

NOTE.– This diffusion coefficient, or diffusivity, measures well the variation of a surface as a function of time, a spreading (m^2/s).

Diffusion by phonons

Here, we consider, for example, longitudinal acoustic phonons.

Vibrations of the perturbation introduce a disturbance capable of causing the electrons (or holes) to pass from one state to another.

To define this perturbation, the so-called deformation potential method is used.

The propagation of a harmonic wave (w, K) causes the appearance of local deformations of the crystal.

Let $u(x,t) = u_0 e^{-i(w\,t - Kr + p/2)}$ be the displacement caused by the elastic wave.

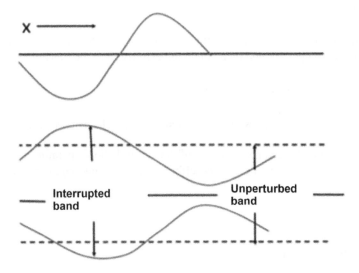

Figure A.2. *"Gap" in the presence of an acoustic wave. For a color version of this figure, see: http://www.iste.co.uk/gontrand/analog.zip*

Deformation u induces a variation in volume:

$$div\ u = K\ u_0\ e^{-i(w\,t - Kr)}$$

[A.36]

Local deformations produced by the wave give rise to a displacement of the edge of the conduction band and of the valence band such that:

$$E_C = E_{CO} + u_D \tag{A.37}$$

The additional potential energy that an electron acquires in a deformed lattice is called the deformation potential; it is related to the magnitude, sometimes denoted as E_1, which is called the deformation potential constant, defined by:

$$U_D(tr) = E_1.\overrightarrow{grad.}\ \vec{u}(r) \tag{A.38}$$

Of course, $\vec{u}(r)$ represents the displacement of the atom, considering all the exited modes of the network (possibly longitudinal modes and transverse modes).

Or:

$$u_r = \sum_\alpha u_{0\alpha}.K_\alpha e^{-i(u_\alpha t - \overrightarrow{K_\alpha}.\vec{r})} \tag{A.39}$$

Knowing the deformation potential, we must calculate the matrix element and the perturbation operator and then obtain the expression of the transition probability.

Let $P = \frac{1}{h}|< k|H|k' >|^2\ D_f(\varepsilon_i)$

where: $|< k|H|k' >| = |< k|U_D|k' >|$ and $D_f(\varepsilon_i)$ is the density of final states.

To calculate the matrix element of the perturbation operator, Block waves are used in the approximation of the effective mass by normalizing to the volume $G = L^3$.

Let us consider

$$\psi_\kappa y_k(r,t) = y_k(r)\ .e^{-i(\frac{2\pi E}{h}t\ -\ \vec{k}.\vec{r})} \tag{A.40}$$

$$< k|u_D|k' >= E_=\frac{E_1 u_0}{G}\ e^{-i(\frac{2\pi}{h}(\varepsilon - \varepsilon' + \frac{h\omega}{2\pi})t}\int_G^{\cdot} e^{i(k+K-k')r}\ dG \tag{A.41}$$

$$= EE_1 u_0 K\ e^{-i(\frac{2\pi}{h}(\varepsilon - \varepsilon' + \frac{h}{2\pi})t}d._{k+K,k'} \tag{A.42}$$

The probability of transition from state k to k' is therefore:

$$w(k,k') = (4p^2/h)(E_1 u_0 K)^2\ d(e + (h/2p)w - e')d_{k+K-k'} \tag{A.43}$$

The expression obtained contains two quantities, the presence of which ensures the conservation of energy and quasi-momentum.

$$E' = E + (h/2p)w$$

And

$$(h/2p) k' = (h/2p)k + (h/2p)K$$

There must therefore be an exchange of energy and quasi-momentum between the electron and the phonon.

Relationships:

$$E' = E + (h/2p).w$$

$$(h/2\pi) k' = (h/2\pi).k \pm (h/2\pi)K$$

generalize those above (case where the exponent of the exponential function is positive or negative) and mean that the energy of the carrier can be increased or decreased by the value of the energy of the phonon.

The result obtained above shows that only the term in $e^{i \vec{k}.\vec{z}}$ will have a nonzero contribution.

Relaxation time

Longitudinal acoustic phonons.

We can write as follows:

$$\overrightarrow{E_1.grad}\ U_k(\vec{r}) = E_1.i\vec{K}l_k (a_k e^{i \vec{k}.\vec{r}} - a_k^*(e^{-i \vec{k}.\vec{r}}) \qquad [A.44]$$

In this expression, the vibrations propagating colinearly to the vector K are considered. These are therefore longitudinal modes.

$$\overrightarrow{u_r} = \overrightarrow{l_K} (a_k e^{i \vec{k}.\vec{r}} - a_k^*(e^{i \vec{k}.\vec{r}}) \qquad [A.45]$$

$\overrightarrow{l_K}$ is a unit vector in direction $\vec{K}$

We see that P is proportional to $E_1^2 |K|^2 |a_k|^2$

Let us evaluate the quantity above.

The energy associated with an excited vibration mode is: $k_bT/2$.

The vibration mode in question is of interest to N atoms of mass M.

The acoustic branches correspond to in-phase vibrations of the atoms of the elementary mesh; the amplitude and the phase of the vibrations of all the atoms are then identical and, in the vicinity of the center of the Brillouin zone, there is a linear relationship between the frequency and the wave number, that is:

$$w = c.K$$

where c is the speed of sound.

Let us calculate the amplitude of the speed.

$$\overrightarrow{u_r} = \overrightarrow{1_K}\left(a_k e^{i\,\overline{k}.\overline{r}} - a_k^*(e^{-i\,\overline{k}.\overline{r}})e^{i\omega t}\right) \tag{A.46}$$

$$u(K,t) = i\omega\left(a_k e^{i\,\overline{k}.\overline{r}}\right) + a_k^*(e^{-i\,\overline{k}.\overline{r}})e^{i\omega t} \tag{A.47}$$

So:

$$[u(K,t)]' = i\omega.(a_k e^{i\,\overline{k}.\overline{r}}) + a_k^* e^{i\,\overline{k}.\overline{r}}).e^{i\omega t} \tag{A.48}$$

It follows that the mean value of the square of the speed is given by:

$$<v_k^2> = 2\,\omega^2\,(a_k)^2 \tag{A.49}$$

From this, we deduce that:

$$\frac{1}{2}NM < v_k^2 > = NM\omega^2\,(a_k)^2 = \frac{kT}{2} \tag{A.50}$$

Or:

$$(a_k)^2 = \frac{kT}{2NV\omega^2} \tag{A.51}$$

If the specific mass of the material is used, the following is deduced:

$$(a_k)^2 = \frac{kT}{2\rho V|K|^2 c^2} \tag{A.52}$$

For a crystal of unit volume, we have:

$$P \sim E_1{}^2 K^2 a_k{}^2 = \frac{kT E_1{}^2}{2\rho c^2} \tag{A.53}$$

Strictly speaking, we obtain:

$$P = \frac{2\pi}{h} \frac{kT E_1{}^2}{2\rho c^2} \frac{d^3k}{4\pi^3} \tag{A.54}$$

This expression shows that P does not depend on the collision angle, which means that these shocks with the network are isotropic.

Let P be calculated as a function of energy.

$$P = \frac{4\pi^2}{h} \frac{kT E_1{}^2}{2\rho c^2} \frac{4\pi k_i{}^2 dk_i}{4\pi^3} \tag{A.55}$$

Let: $\dfrac{dE}{\tau(E)} = \dfrac{4\pi^2}{h} \dfrac{kT E_1{}^2}{2\rho c^2} \dfrac{4\pi k_i{}^2 dk_i}{4\pi^3}$ \hfill (A.56)

However:

$$\frac{4\pi k_i{}^2 dk_i}{4\pi^3} = n(E)dE = \frac{4\pi^2}{h^3}\left((2m)^{\frac{3}{2}} (E)^{\frac{1}{2}} dE \right) \tag{A.57}$$

$4\pi k_i{}^2 dk_i$ is the volume of the shell between the radii k and k + dk/state density.

Or:

$$\frac{1}{\tau(E)} = \frac{8\pi^3}{h^4} \frac{kT E_1{}^2}{2\rho c^2} (2m)^{3/2} (E)^{1/2} \tag{A.58}$$

The expression for the relaxation time is deduced therefrom:

$$\frac{1}{\tau(E)} = \int_0^\pi \frac{1}{\tau}(1 - \cos\theta)\sin\theta\, d\theta \tag{A.59}$$

$$= \frac{1}{\tau} = \frac{8\pi^3}{h^4} \frac{kT E_1{}^2}{\rho c^2} (2m)^{\frac{3}{2}} (E)^{\frac{1}{2}} \tag{A.60}$$

Of course, if the semiconductor is degenerate:

$$t = t_s \sim E_F{}^{-1/2} kT^{-1} \tag{A.61}$$

If the semiconductor is nondegenerate:

$$<t_s> = \frac{h^4}{8\pi^3 kT E_1^2} \frac{\rho c^2}{(2m)^{3/2}} \frac{\int_0^\infty E. \exp^{-E/kT} dE}{\int_0^\infty E^{3/2}. \exp^{-E/kT} dE} \qquad [A.62]$$

We deduce from this the expression of mobility:

$$\mu = \frac{e<\tau_\sigma>}{m} = \frac{(2m)^{1/2} h^4}{3kT E_1^2 (kT)^{3/2}} \cdot \frac{\rho c^2 e}{(m)^{5/2}} \qquad [A.63]$$

NOTE.–

A law of variation of the mobilized at $T^{-3/2}$ is observed. We can recall that for I.I, the law of variation is in $T^{+3/2}$. Thus, the bell curve is generally observed for nondegenerate semiconductors.

1) The expression of the value of m introduced here corresponds to the mass of density of states. In the expression $\frac{e<\tau_\sigma>}{m}$, the value of m represents the conduction mass.

Equality between these two masses exists only in the simple case of a parabolic and isotropic band.

2) In the case of diffusion by optical modes, it is essential to take account of $(h/2p).w$ since the energy of the optical vibrations is high. As at low temperature, the number of optical phonons must be small and cannot be replaced u_0^2 by kT, since u_0^2 is proportional to the number of optic phonons:

$$N = \frac{1}{e^{\left(\frac{h\omega}{2\pi}\right)}} - 1 \qquad [A.64]$$

At high temperatures, the number of phonons will be proportional to kT as in the case of acoustic phonons, with the difference that T is now independent of the energy of the charge carriers.

We have just presented a calculation concerning parabolic bands; this is the case of Si-p but also of Si-n but then the effective masses depend on the crystallographic direction. It is then no longer necessary to deal with spheres with regard to the dispersion relationship e(k), but with ellipsoids that are dependent on the transverse and longitudinal masses with respect to the direction <100>. In particular, if the Boltzmann equation is solved by numerical methods, these relationships can be "made into a sphere" by a change of appropriate variable involving m_{111} and m_{100}.

Let us take the case of the relationship e(k) of the type $e(k) = (\frac{h}{2\pi})^2 \frac{k^2}{2m} = e(k)$ $(1+ae(k))$.

NOTE.– It can be seen that, at low energy, the energy is proportional to k^2; for high energies, the latter vary linearly in k.

We get:

$$<t> \; = \; \frac{V}{8\pi^3} \int_0^\infty \frac{4\pi}{h} \cdot \frac{kT\,E_1}{\rho c V}^2 \cdot \left|\vec{k} - \vec{k'}\right| . G(k,k')Na.\, d^3k \qquad [A.65]$$

$$\text{with: } Na \; = \; \frac{2\,\pi kT}{.\left|\vec{k} - \vec{k'}\right| hc}\, s \qquad [A.66]$$

$$G(kk') = (\frac{1+\alpha\varepsilon + \alpha\varepsilon\cos\theta'}{1+2\alpha\varepsilon})^{\,2} \qquad [A.67]$$

$$<t> \; = \; \frac{V}{8\pi^3} \int_0^\infty \frac{4\pi}{h} \cdot \frac{kT\,E_1}{\rho c V}^2 \cdot \left|\vec{k} - \vec{k'}\right| .\; k'^2 \sin\theta'\; d\theta'\; d\varphi' \delta\,(\varepsilon - \varepsilon') \qquad [A.68]$$

$$<t> \; = \; \frac{V}{8\pi^3} \int_0^\infty \frac{4\pi}{h} \cdot \frac{kT\,E_1}{\rho c V}^2 \cdot \left|\vec{k} - \vec{k'}\right| .\; k'^2 \sin\theta'\; d\theta'\; d\varphi' \delta\,(\varepsilon - \varepsilon') \qquad [A.69]$$

$$\frac{V}{8\pi^3} \frac{4\pi}{h} \frac{kT\,E_1}{\rho c V}^2 \cdot \frac{k^2}{\frac{\partial \varepsilon}{\partial k}} \int_0^\infty 1 + \alpha\varepsilon + \alpha\varepsilon\cos\theta'.\; k^2 \sin\theta'\; \delta\; d\varphi' \qquad [A.70]$$

But, for AsGa-type non- – parabolic bands: $(\frac{h}{2\pi})^2 \frac{k^2}{2m} = e(k)\; (1+ae(k))$ ➔

$$\frac{\partial \varepsilon}{\partial k} \; = \; (\frac{h}{2\pi})^2 \frac{k^2}{1+2\alpha\varepsilon}) \qquad [A.71]$$

$$\frac{V}{8\pi^3} \frac{4\pi}{h} \frac{kT\,E_1}{\rho c V}^2 \cdot \frac{k^2}{\frac{\partial \varepsilon}{\partial k}} \int_0^\infty 1 + \alpha\varepsilon + \alpha\varepsilon\cos\theta'.\; k^2\sin\theta'\; d\theta'\; d\varphi' \qquad [A.72]$$

$V\frac{4\pi}{h} \frac{kT\,E_1}{\rho c V}^2 \frac{.m}{1+2\alpha\varepsilon}. \, 2(1 + \alpha\varepsilon)^2 + (\alpha\varepsilon)^2 \int_0^\pi \cos\theta'^2 \sin\theta' d\,\theta' + 2(1 + \alpha\varepsilon\,)\alpha\varepsilon \int_0^\pi \cos\theta' \sin\theta'\; d\theta'.$

$\cos q' = u$ ➔ $du = - \sin q' d\,q'$

$$\int_0^\pi \cos\theta' \sin\theta' d\,\theta' = [\frac{u^2}{2}]^1_{-1} = 0$$

$$\int_0^\pi \cos\theta'^2 \sin\theta' d\,\theta' = [\frac{u^3}{3}]^1_{-1} = \frac{2}{3}$$

Or $(\frac{h}{2\pi})^2 \frac{k^2}{2m} = e(k)(1+a\ e\ (k)\ ) = g$ (a: non-parabolic factor: eV^{-1})

So, $k = g^{1/2}.\frac{2hm^{1/2}}{2\pi}$

Here:

$$\frac{1}{\tau_{ac}} = \frac{(2m)^{3/2}}{2\pi} \frac{kT\ E_1^{\ 2}}{\rho c^2.(\frac{h}{2\pi})^4} \frac{(\gamma)^{1/2}}{(1+2\alpha\varepsilon)}.((1+\alpha\varepsilon)^2 + 1/3(\alpha\varepsilon)^2) \qquad [A.73]$$

These calculations are relatively tedious, but do not present any real difficulties. What are worse, for example, are interactions, polar optics, piezoelectric, etc.

We will show the calculation for polar optics: then we will go to the essentials for the other two.

– Polar optical interactions

$$-<1> = \frac{\gamma^{1/2}}{4\pi h\varepsilon_0} \int_0^\pi \int_0^\infty 2\pi. \frac{1}{4\pi|\vec{k}-\vec{k'}|^2} (\frac{1}{\varepsilon_\infty} - \frac{1}{\varepsilon_0}) G(\vec{k},\vec{k'}).\left(N_0 + \frac{1}{2} \pm \frac{1}{2}\right) \qquad [A.74]$$

$$\delta\ ((\varepsilon(k') - \varepsilon(k) \pm (\frac{h\omega}{2\pi})k^2 \sin \beta' d\ \beta' dk'$$

After a few calculations, we get:

$$-<1> = \frac{V\omega_0\ e^2 m^{1/2}}{2^{1/2} 4\pi h\varepsilon_0} (\frac{1}{\varepsilon_\infty} - \frac{1}{\varepsilon_0}) \frac{\gamma^{1/2}}{1+2\alpha\varepsilon}$$

$$* \int_0^\pi \int_0^\infty 2\pi. \frac{(1+\alpha\varepsilon)^{1/2}(1+\alpha\varepsilon')^{1/2} + \alpha\varepsilon^{1/2}\varepsilon'^{1/2}}{\gamma+\gamma' - 2\gamma^{1/2}\gamma'^{1/2}\cos \beta'} \sin \beta' d\ \beta'.\left(N_0 + \frac{1}{2} \pm \frac{1}{2}\right). \quad [A.75]$$

To calculate, we have:

$$I = \int_0^\beta \frac{((1+\alpha\varepsilon)^{1/2}(1+\alpha\varepsilon')^{1/2} + \alpha\varepsilon^{1/2}\varepsilon'^{1/2}\cos \beta')^2}{\gamma+\gamma' - 2\gamma^{1/2}\gamma'^{1/2}\cos \beta'} \sin \beta' d\ \beta'$$

$A = (1 + \alpha\varepsilon)^{1/2}(1 + \alpha\varepsilon')^{1/2}$ $\qquad\qquad C = \gamma + \gamma'$

$B = \alpha\varepsilon^{1/2}\varepsilon'^{1/2}$ $\qquad\qquad\qquad\qquad D = 2\gamma^{1/2}\gamma'^{1/2}$

$$I = \int_0^\beta \frac{(A+B\cos \beta')^2}{C-D\cos \beta'} \sin \beta' d\ \beta'$$

$C - D\cos \beta' = X ==> D\sin \beta' = dX$

$$I_1 = \frac{A^2}{D} \int_{C-D}^{D\cos\beta} \frac{dX}{X}$$

$$I_1 = \frac{A^2}{D} \text{Log} \left| \frac{C - D\cos\beta}{C-D} \right|$$

$$I_2 = 2AB \int_1^{\cos\beta} \frac{XdX}{C - DX}$$

$$Y = C - DX \rightarrow dX = -\frac{1}{D}dY$$

$$I_2 = \frac{2AB}{D} \cdot \int_{C-D}^{\cos\beta} \frac{\frac{C-Y}{D}dY}{Y}$$

Here:

$$I_2 = \frac{2AB}{D}\cos\beta - \frac{2AB}{D} + \frac{2ABC}{D} \cdot \text{Log} \left| \frac{C - D\cos\beta'}{C-D} \right|$$

$$I_3 = \frac{B^2}{D^3} \cdot \int_{C-D}^{C-D\cos\beta} \frac{(C-Y)^2 \, dY}{Y}$$

In fine:

$$I_3 = \frac{B^2}{D^3} \cdot \cos^2\beta + \frac{B^2 C}{D^2} \cdot \cos\beta - \frac{B^2 C}{D^2} - \frac{B^2 C}{2D} - \frac{B^2 C^2}{D^3} \text{Log} \left| \frac{C - D\cos\beta}{C-D} \right|$$

Since $I = I_1 + I_2 + I_3$.

$$I = A_2 \cos^2\beta + A_1 \cdot \cos\beta + A_0 + A' \text{ Log} \left| \frac{C - D\cos\beta}{C-D} \right| \tag{A.76}$$

$$A_2 = \frac{B^2}{2D}$$

$$A_1 = \frac{2AB}{D} + \frac{B^2 C}{D^2}$$

$$A_0 = -\left(\frac{2AB}{D} + \frac{B^2 C}{D^2} + \frac{B^2}{2D} \right)$$

$$A' = \frac{1}{D}\left(A + +\frac{BC}{D} \right)^2$$

$$I(\beta = \pi) = \frac{1}{C'} \cdot \left(B + A. \text{ Log} \left| \frac{C - D\cos\beta}{C-D} \right| \right)$$

$$<1/t> = \frac{P^2}{4\pi^2} \frac{2\pi}{h} \frac{kT e^2}{\varepsilon_0 \varepsilon_s} \int_a^\beta \frac{(1+\alpha\varepsilon+\alpha\varepsilon\cos\theta)^2}{(1+2\alpha\varepsilon)^2} \cdot \frac{1}{|\vec{k}-\vec{k'}|^2} \cdot \delta(\varepsilon-\varepsilon')d^3k'$$

$$= \frac{1}{8\pi^3} \frac{4\pi}{h} \frac{kT E_1^2}{\rho cV} \frac{.1}{\frac{\partial\varepsilon}{\partial k}} \int_a^\beta \frac{(1+\alpha\varepsilon+\alpha\varepsilon\cos\theta)^2}{1-\cos\beta} k^2 \sin\beta' \, d\beta' \, d\varphi'$$

In fine:

$$<1/t> = \frac{m}{8\pi h^3} \frac{kT P^2}{k} \frac{.1}{1+2\alpha\varepsilon}[(1-\cos\beta)^2 - (1-\cos a)^2](\alpha\varepsilon)^2 -$$
$$(1-\cos\beta)(1-\cos a).2(\alpha\varepsilon(1+2\alpha\varepsilon)$$

$$\text{Log}\left|\frac{1-\cos\beta}{1-\cos a}\right|((1+2\alpha\varepsilon)^2 \qquad\qquad\qquad [A.77]$$

For b = p:

$$1/t(k,kp) = \frac{m}{8\pi h^3} \frac{kT P^2}{k} \frac{.1}{1+2\alpha\varepsilon}[4 - (1-\cos a)^2](\alpha\varepsilon)^2 - 2(1-\cos a).$$
$$2(\alpha\varepsilon(1+2\alpha\varepsilon) \qquad\qquad\qquad\qquad [A.78]$$

$$\text{Log}\left|\frac{2}{1-\cos a}\right|((1+2\alpha\varepsilon)^2$$

If $(\vec{k},\vec{k'})$ represents the probability per unit time that a carrier in state $\vec{k}$ s will be transferred to another state for the ith type of collision:

$$<\lambda_i> = \int S_i(\vec{k},\vec{k'}) \, d^3k' \qquad\qquad\qquad [A.79]$$

Intervalley acoustic phonons

a) Elastic collision hypothesis

If $(h/2p).s. |\vec{k}-\vec{k'}|^2 << e$ and if we consider an isotropic effective mass.

Going from

$$S_i(\vec{k},\vec{k'}) = \frac{4\pi^2}{h} \frac{kT Z_a^2}{\rho s^2 V} \cdot \delta(\varepsilon(k)-\varepsilon(k'))$$

with: $\varepsilon(k) = \varepsilon(k')$

Number of energy phonons $(h/2p)s|\vec{K} - \vec{k'}|$

$$\bar{n}_a \sim \frac{2\pi}{hs|\vec{K} - \vec{k'}|} \frac{kT}{\rho s^2 V} \qquad [A.80]$$

We can deduce that:

$$\lambda_i(\vec{k}) = \frac{4\pi^2}{h} \frac{kT\,Z_a{}^2}{\rho s^2 V} \iiint . \; k'^2 \sin\theta'\, d\theta'\, d\varphi'\, d(\varepsilon(\vec{k'}) - \varepsilon(\vec{k})) \qquad [A.81]$$

$p \geq q \geq 0$ and $p \geq y \geq 0$

Finally:

$$\frac{1}{\tau_{ac}} = \frac{8\pi(2m)^{3/2}\, kT\, Z_a{}^2}{\rho s^2 h^4} . (\varepsilon(1 + \alpha\varepsilon)^{21/2}(1 + 2\alpha\varepsilon) \qquad [A.82]$$

b) Elastic hypothesis and anisotropic effective mass.

The same formula is used by replacing the effective mass of state density.

$M_D = (m_l\, m_l\, m_l)1/3$

Typically, for Si-N:

$Za = 9$ eV

$r = 2.328$ g/cm^3

$s = 9.02$ m/s

$a = 0.5$ eV $^{-1}$

$m_l = 0.9163\, m_0$

$m_t = 0.1905\, m_0$

1) Inelastic interval transitions:

The energy variation in a collision, bringing the carrier from a valley i to a valley j, is constant set equal to $(h/2p).w_{ij}$.

$$\bar{n}_{ij} = \frac{1}{e\left(\frac{h\omega_{ij}}{kT}\right)} - 1 \qquad [A.83]$$

$$S_{ij}(\vec{k}, \vec{k'}) = \frac{4\pi^2}{h} . \frac{kT\, Z_{ij}{}^2}{2\rho\omega_{ij} V} \bar{n}_{ij} \delta(\varepsilon(\vec{k'}) - \varepsilon(\vec{k}) - \frac{h\omega_{ij}}{2\pi}) \qquad [A.84]$$

For absorption:

$$S_{ij}(\vec{k}, \vec{k'}) = \frac{4\pi^2}{h} \frac{kT\, Z_{ij}^2}{2\rho\omega_{ij}V} \bar{n}_{ij}\delta(\varepsilon(\vec{k'}) - \varepsilon(\vec{k}) + \frac{h\omega_{ij}}{2\pi})$$ [A.85]

Then:

$$S_{ij}(\vec{k}, \vec{k'}) = \frac{4\pi^2}{h} \frac{kT\, Z_{ij}^2}{2\rho\omega_{ij}V} \bar{n}_{ij}\delta(\varepsilon(\vec{k'}) - \varepsilon(\vec{k}) + \frac{h\omega_{ij}}{2\pi})$$ [A.86]

Note that $S_{ij}(\vec{k}, \vec{k'}) = 0$ si $e < (h/2p)w_{ij}$.

A.5. Influence of the magnetic field on the movement of electrons

The Lorentz force is used:

$$\frac{h}{2\pi}\frac{\partial\vec{k}}{\partial t} = -q\vec{E} - \vec{v}\wedge\vec{B} \quad (\frac{d\vec{p}}{dt} = \sum external\ forces° ;$$ [A.87]

$$\vec{p} = \frac{h\vec{k}}{2\pi}\ \text{and}\ \vec{v}\wedge\vec{B}\ \text{is homogenous to an electric field)}$$

with

$$v = c$$

$$v = \frac{hk}{m^*}\ \text{(parabolic bands)} \qquad\qquad \frac{\partial\vec{k}}{\partial t} = m^*\frac{d\vec{v}}{dt}$$

$$m^*\frac{d\vec{v}}{dt} = -q\vec{E} - q\vec{v}\wedge\vec{B}$$

$\vec{E}$ belongs to the plane $<x,y>$; $\vec{B}$ is on z;

$$m^*\frac{dv_x}{dt} = -qE_x - qv_yB$$ [A.88]

$$m^*\frac{dv_y}{dt} = -qE_y + qv_xB$$ [A.89]

$$m^*\frac{dv_z}{dt} = 0$$

If we write: $V = v_x + iv_y$

$$V = v_x + iv_y$$

So:

$$\frac{dV}{dt} = \frac{-q}{m^*}\left(E_x + iE_y\right) - \omega(v_y - iv_x) \qquad \omega = \frac{qB}{m^*} \qquad [A.90]$$

$$\frac{dV}{dt} = \frac{-q}{m^*}\left(E_x + iE_y\right) - i\omega(-iv_y + v_y) \qquad [A.91]$$

$$\frac{dV}{dt} - i\omega V = \frac{-q}{m^*}\left(E_x + iE_y\right) \qquad [A.92]$$

Solution:

$$V = Ke^{i\omega t} + \frac{q}{m^* i\omega}\left(E_x + iE_y\right) \qquad [A.93]$$

A t = 0, V = V$_0$

$$V_0 = K + \frac{-q}{m^* i\omega}\left(E_x + iE_y\right) \qquad [A.94]$$

Therefore:

$$K = V_0 - \frac{q}{m^* i\omega}\left(E_x + iE_y\right)$$

$$cl: V = V_0 e^{i\omega t} + \frac{-q}{m^* i\omega}\left(E_x + iE_y\right)e^{i\omega t} + \frac{q}{m^* i\omega}\left(E_x + iE_y\right) \qquad [A.95]$$

$$V = V_0 e^{i\omega t} - \frac{q}{m^* i\omega}\left(E_x + iE_y\right).\frac{(e^{i\omega t} - 1)}{i\omega} \qquad [A.96]$$

The probability that an electron has not collided at time t is given via $e^{-t/\tau}$.

t is the mean time between collisions.

$$\int_0^\infty e^{-\frac{t}{\tau}}dt = \tau$$

$$\frac{1}{\tau}\int_0^\infty e^{-t/\tau}dt = 1$$

Averaging of V on all collisions.

$$V = \frac{1}{\tau}\int_0^\infty V(t)\, e^{-t/\tau}dt \qquad [A.97]$$

$$V = V_0 \frac{1+i\omega\tau}{1+\omega^2\tau^2} - \frac{q(E_x+iE_x)\tau}{m^*}\frac{1+i\omega\tau}{1+\omega^2\tau^2} \qquad [A.98]$$

References

Bloch (1929). Über die Quantenmechanik der Elektronen in Kristallgittern. *Zeitschrift für Physik*, 52(7–8), 555–600.

Buffat, M. (2007). Méthode des éléments finis en Mécanique. Approche illustrée avec Maple et Matlab. Université Claude Bernard, Lyon [Online]. Available at: https://perso.univ-lyon1.fr/marc.buffat/COURS /LIVREEF_HTML/.

Castagné, R. and Vapaille, A. (1987). *Dispositifs et circuits intégrés semi-conducteurs : physique et technologie*. Dunod, Paris.

Gontrand C. (2018). *Micro-nanoelectronics Devices*. ISTE Press Ltd, London, and Elsevier, Amsterdam.

ISE-TCAD Manuals (2002). Release 8.0, Integrated Systems Engineering. ETH Zurich.

Moore, G.E. (1965). Cramming more components onto integrated circuit. *Electronics*, 38(8), 4.

Nag, B.R. (1980). *Electron Transport in Compound Semiconductors*. Springer-Verlag, Berlin Heidelberg, New York.

Nougier, J.-P. (1991). *Méthodes de calcul numérique*, 3rd edition. Elsevier Masson, Issy-les-Moulineaux.

SILVACO (n.d.). SILVACO [Online]. Available at: https://silvaco.com.

Warner, R.M. (1968). *Circuits intégrés*. Dunod, Paris.

Wilson, A.H. (1931). The theory of electronic semi-conductors II. *Proc. R. Soc. Lond. A*, 134(823), 277–287.

Index

Other titles from

in

Electronics Engineering

2023

COURRIER Thierry, QUARTIER Laurent
Ondes Martenot with Tubes

NGUYEN Thien-Phap
Defects in Organic Semiconductors and Devices

2022

METHAPETTYPARAMBU PURUSHOTHAMA Jayakrishnan, PERRET Etienne,
VENA Arnaud
*Non-Volatile CBRAM/MIM Switching Technology for Electronically
Reconfigurable Passive Microwave Devices: Theory and Methods for
Application in Rewritable Chipless RFID*

2021

NGUYEN Thien-Phap
Organic Electronics 1: Materials and Physical Processes
Organic Electronics 2: Applications and Marketing

2020

CAPPY Alain
Neuro-inspired Information Processing

2014

APPRIOU Alain
Uncertainty Theories and Multisensor Data Fusion

CONSONNI Vincent, FEUILLET Guy
Wide Band Gap Semiconductor Nanowires 1: Low-Dimensionality Effects and Growth
Wide Band Gap Semiconductor Nanowires 2: Heterostructures and Optoelectronic Devices

GAUTIER Jean-Luc
Design of Microwave Active Devices

LACAZE Pierre Camille, LACROIX Jean-Christophe
Non-volatile Memories

TEMPLIER François
OLED Microdisplays: Technology and Applications

THOMAS Jean-Hugh, YAAKOUBI Nourdin
New Sensors and Processing Chain

2013

COSTA François, GAUTIER Cyrille, LABOURE Eric, REVOL Bertrand
Electromagnetic Compatibility in Power Electronics

KORDON Fabrice, HUGUES Jérôme, CANALS Agusti, DOHET Alain
Embedded Systems: Analysis and Modeling with SysML, UML and AADL

LE TIEC Yannick
Chemistry in Microelectronics

2012

BECHERRAWY Tamer
Electromagnetism: Maxwell Equations, Wave Propagation and Emission

BELLEVILLE Marc, CONDEMINE Cyril
Energy Autonomous Micro and Nano Systems

CLAVERIE Alain
Transmission Electron Microscopy in Micro-nanoelectronics

LALAUZE René
Chemical Sensors and Biosensors

Printed and bound by CPI Group (UK) Ltd, Croydon, CR0 4YY

16/04/2025

14658458-0002